SpringerBriefs in Mathematics

SpringerBriefs present concise summaries of cutting-edge research and practical applications across a wide spectrum of fields. Featuring compact volumes of 50 to 125 pages, the series covers a range of content from professional to academic. Briefs are characterized by fast, global electronic dissemination, standard publishing contracts, standardized manuscript preparation and formatting guidelines, and expedited production schedules.

Typical topics might include:

A timely report of state-of-the art techniques A bridge between new research results, as published in journal articles, and a contextual literature review A snapshot of a hot or emerging topic An in-depth case study A presentation of core concepts that students must understand in order to make independent contributions

SpringerBriefs in Mathematics showcases expositions in all areas of mathematics and applied mathematics. Manuscripts presenting new results or a single new result in a classical field, new field, or an emerging topic, applications, or bridges between new results and already published works, are encouraged. The series is intended for mathematicians and applied mathematicians. All works are peer-reviewed to meet the highest standards of scientific literature.

Titles from this series are indexed by Scopus, Web of Science, Mathematical Reviews, and zbMATH.

More information about this series at https://link.springer.com/bookseries/10030

Matteo Cicuttin · Alexandre Ern · Nicolas Pignet

Hybrid High-Order Methods

A Primer with Applications to Solid Mechanics

Springer

Matteo Cicuttin
Department of Electrical Engineering
and Computer Science, Montefiore
Institute B28
University of Liège
Liège, Belgium

Alexandre Ern
CERMICS
Ecole des Ponts and INRIA Paris
Marne la Vallée and Paris, France

Nicolas Pignet
ERMES
Électricité de France R&D
Palaiseau, France

ISSN 2191-8198 ISSN 2191-8201 (electronic)
SpringerBriefs in Mathematics
ISBN 978-3-030-81476-2 ISBN 978-3-030-81477-9 (eBook)
https://doi.org/10.1007/978-3-030-81477-9

Mathematics Subject Classification: 65N12, 65M12, 74S05, 65N30, 74B05, 74B20, 74C15, 74M10, 74M15, 74J05, 65-04

This Springer imprint is published by the registered company Springer Nature Switzerland AG
The registered company address is: Gewerbestrasse 11, 6330 Cham, Switzerland

Preface

Hybrid High-Order (HHO) methods attach discrete unknowns to the cells and to the faces of the mesh. At the heart of their devising lie two intuitive ideas: (i) a local operator reconstructing in every mesh cell a gradient (and possibly a potential for the gradient) from the local cell and face unknowns and (ii) a local stabilization operator weakly enforcing in every mesh cell the matching of the trace of the cell unknowns with the face unknowns. These two local operators are then combined into a local discrete bilinear form, and the global problem is assembled cellwise as in standard finite element methods. HHO methods offer many attractive features: support of polyhedral meshes, optimal convergence rates, local conservation principles, a dimension-independent formulation, and robustness in various regimes (e.g., no volume-locking in linear elasticity). Moreover, their computational efficiency hinges on the possibility of locally eliminating the cell unknowns by static condensation, leading to a global transmission problem coupling only the face unknowns.

HHO methods were introduced in [77, 79] for linear diffusion and quasi-incompressible linear elasticity. A high-order method in mixed form sharing the same devising principles was introduced in [78], and shown in [6] to lead after hybridization to a HHO method with a slightly different, yet equivalent, writing of the stabilization. The realm of applications of HHO methods has been substantially expanded over the last few years. Developments in solid mechanics include nonlinear elasticity [26], hyperelasticity [1], plasticity [2, 3], poroelasticity [16, 27], Kirchhoff–Love plates [19], the Signorini [44], obstacle [59] and two-membrane contact [69] problems, Tresca friction [53], and acoustic and elastic wave propagation [33, 34]. Those related to fluid mechanics include convection-diffusion in various regimes [74], Stokes [6, 81], Navier–Stokes [23, 45, 82], Bingham [43], creeping non-Newtonian [24], and Brinkman [22] flows, flows in fractured porous media [47, 106], single-phase miscible flows [7], and elliptic [35] and Stokes [32] interface problems. Other interesting applications include the Cahn–Hilliard problem [49], Leray–Lions equations [72], elliptic multiscale problems [60], H^{-1} loads [95], spectral problems [38, 41], domains with curved boundary [21, 35, 36], and magnetostatics [48].

Bridges and unifying viewpoints emerged progressively between HHO methods and several other discretization methods which also attach unknowns to the mesh cells and faces. Already in the seminal work [79], a connection was established between the lowest-order HHO method and the hybrid finite volume method from [97] (and, thus, to the broader setting of hybrid mimetic mixed methods in [85]). Perhaps the most salient connection was made in [62] where HHO methods were embedded into the broad setting of Hybridizable Discontinuous Galerkin (HDG) methods [64]. One originality of equal-order HHO methods is the use of the (potential) reconstruction operator in the stabilization. Moreover, the analyses of HHO and HDG methods follow somewhat different paths, since the former relies on orthogonal projections, whereas the latter often invokes a more specific approximation operator [65]. We believe that the links between HHO and HDG methods are mutually beneficial, as, for instance, recent HHO developments can be transposed to the HDG setting. Weak Galerkin (WG) methods [148, 149], which were embedded into the HDG setting in [61 Sect. 6.6], are, thus, also closely related to HHO. WG and HHO were developed independently and share a common devising viewpoint combining reconstruction (called weak gradient in WG) and stabilization. Yet, the WG stabilization often relies on plain least-squares penalties, whereas the more sophisticated HHO stabilization is key to a higher-order consistency property. Furthermore, the work [62] also bridged HHO methods to the nonconforming virtual element method [10, 119]. Finally, the connection to the multiscale hybrid mixed method from [105] was uncovered in [46].

A detailed monograph on HHO methods appeared this year [73]. The present text is shorter and does not cover as many aspects of the analysis and applications of HHO methods. Its originality lies in targetting the material to computational mechanics without sacrificing mathematical rigor, while including on the one hand some mathematical results with their own specific twist and on the other hand numerical illustrations drawn from industrial examples. Moreover, several topics not covered in [73] are treated here: domains with curved boundary, hyperelasticity, plasticity, contact, friction, and wave propagation. The present material is organized into eight chapters: the first three gently introduce the basic principles of HHO methods on a linear diffusion problem, the following four present various challenging applications to solid mechanics, and the last one reviews implementation aspects.

This book is primarily intended for graduate students, researchers (in applied mathematics, numerical analysis, and computational mechanics), and engineers working in related fields of application. Basic knowledge of the devising and analysis of finite element methods is assumed. Special effort was made to streamline the presentation so as to pinpoint the essential ideas, address key mathematical aspects, present examples, and provide bibliographic pointers. This book can also be used as a support for lectures. As a matter of fact, its idea originated from a series of lectures given by one of the authors during the Workshop on Computational Modeling and Numerical Analysis (Petrópolis, Brasil, 2019).

We are thankful to many colleagues for stimulating discussions at various occasions. Special thanks go to G. Delay (Sorbonne University) and S. Lemaire (INRIA) for their careful reading of parts of this manuscript.

Namur, Belgium — Matteo Cicuttin
Paris, France — Alexandre Ern
December 2020 — Nicolas Pignet

Contents

Chapter 1
Getting Started: Linear Diffusion

The objective of this chapter is to gently introduce the hybrid high-order (HHO) method on one of the simplest model problems: the Poisson problem with homogeneous Dirichlet boundary conditions. Our goal is to present the key ideas underlying the devising of the method and state its main properties (most of them without proof). The keywords of this chapter are cell and face unknowns, local reconstruction and stabilization operators, elementwise assembly, static condensation, energy minimization, and equilibrated fluxes.

1.1 Model Problem

Let Ω be an open, bounded, connected, Lipschitz subset of $\mathbb{R}^d$ in space dimension $d \geq 2$. The one-dimensional case $d = 1$ can also be covered, and we refer the reader to Sect. 1.6 for an outline of HHO methods in this setting. Vectors in $\mathbb{R}^d$ and vector-valued functions are denoted in bold font, $\boldsymbol{a}\cdot\boldsymbol{b}$ denotes the Euclidean inner product between two vectors $\boldsymbol{a}, \boldsymbol{b} \in \mathbb{R}^d$ and $\|\cdot\|_{\ell^2}$ the Euclidean norm in $\mathbb{R}^d$. Moreover, $\#S$ denotes the cardinality of a finite set S.

We use standard notation for the Lebesgue and Sobolev spaces; see, e.g., [30, Chaps. 4 and 8], [92, Chaps. 1–4], and [5, 96]. In particular, $L^2(\Omega)$ is the Lebesgue space composed of square-integrable functions over Ω, and $H^1(\Omega)$ is the Sobolev space composed of those functions in $L^2(\Omega)$ whose (weak) partial derivatives are square-integrable functions over Ω. Moreover, $H_0^1(\Omega)$ is the subspace of $H^1(\Omega)$ composed of functions with zero trace on the boundary $\partial\Omega$. Inner products and norms in these spaces are denoted by $(\cdot,\cdot)_{L^2(\Omega)}$, $\|\cdot\|_{L^2(\Omega)}$, $(\cdot,\cdot)_{H^1(\Omega)}$, and $\|\cdot\|_{H^1(\Omega)}$. Recall that for a real-valued function v:

M. Cicuttin et al., *Hybrid High-Order Methods*,
SpringerBriefs in Mathematics,
https://doi.org/10.1007/978-3-030-81477-9_1

$$\|v\|^2_{L^2(\Omega)} := \int_\Omega v^2 \mathrm{d}x, \qquad \|v\|^2_{H^1(\Omega)} := \|v\|^2_{L^2(\Omega)} + \ell_\Omega^2 \|\nabla v\|^2_{L^2(\Omega)}, \tag{1.1}$$

where the length scale $\ell_\Omega :=$ diam (Ω) (the diameter of Ω) is introduced to be dimensionally consistent. Owing to the Poincaré–Steklov inequality (a.k.a. Poincaré inequality; see [92, Remark 3.32] for a discussion on the terminology), there is $C_{\text{PS}} > 0$ such that $C_{\text{PS}}\|v\|_{L^2(\Omega)} \le \ell_\Omega \|\nabla v\|_{L^2(\Omega)}$ for all $v \in H^1_0(\Omega)$.

The model problem we want to approximate in this chapter is the Poisson problem with source term $f \in L^2(\Omega)$ and homogeneous Dirichlet boundary conditions, i.e., $-\Delta u = f$ in Ω and $u = 0$ on $\partial\Omega$. The weak formulation of this problem reads as follows: Seek $u \in V := H^1_0(\Omega)$ such that

$$a(u, w) = \ell(w), \quad \forall w \in V, \tag{1.2}$$

with the following bounded bilinear and linear forms:

$$a(v, w) := (\nabla v, \nabla w)_{\boldsymbol{L}^2(\Omega)}, \qquad \ell(w) := (f, w)_{L^2(\Omega)}, \tag{1.3}$$

for all $v, w \in V$. Since we have $a(v, v) = \|\nabla v\|^2_{\boldsymbol{L}^2(\Omega)}$, the Poincaré–Steklov inequality implies that the bilinear form a is coercive on V. Hence, the model problem (1.2) is well-posed owing to the Lax–Milgram lemma.

1.2 Discrete Setting

In this section, we present the setting to formulate the HHO discretization of the model problem (1.2).

1.2.1 The Mesh

For simplicity, we assume in what follows that the domain Ω is a polyhedron in $\mathbb{R}^d$, so that its boundary is composed of a finite union of portions of affine hyperplanes with mutually disjoint interiors. The case of domains with a curved boundary is discussed in Sect. 3.2.2.

Since Ω is a polyhedron, it can be covered exactly by a mesh $\mathcal{T}$ composed of a finite collection of (open) polyhedral mesh cells T, all mutually disjoint, i.e., we have $\overline{\Omega} = \bigcup_{T\in\mathcal{T}} \overline{T}$. Notice that by definition of a polyhedron, the mesh cells have straight edges if $d = 2$ and planar faces if $d = 3$. For a generic mesh cell $T \in \mathcal{T}$, its boundary is denoted by ∂T, its unit outward normal by $\boldsymbol{n}_T$, and its diameter by h_T. The mesh size is defined as the largest cell diameter in the mesh and is denoted by $h_{\mathcal{T}}$, and more simply by h when there is no ambiguity. When establishing error estimates, one is interested in the process $h \to 0$ corresponding to a sequence of successively

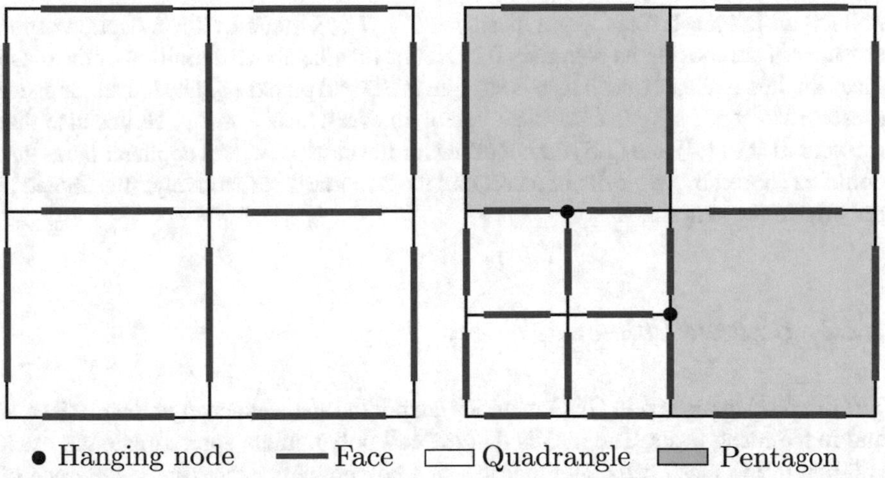

Fig. 1.1 Local refinement of a quadrilateral mesh; the mesh cells containing hanging nodes are treated as polygons (here, pentagons)

refined meshes. In this case, one needs to introduce a notion of shape-regularity for the mesh sequence. This notion is detailed in Sect. 2.1.

The possibility of handling meshes composed of polyhedral mesh cells is an attractive feature of HHO methods. For instance, it allows one to treat quite naturally the presence of hanging nodes arising from local mesh refinement; see Fig. 1.1 for an illustration. However, the reader can assume for simplicity that the mesh is composed of cells with a single shape, such as simplices (triangles in 2D, tetrahedra in 3D) or (rectangular) cuboids, without loosing anything essential in the understanding of the devising and analysis of HHO methods.

Besides the mesh cells, the mesh faces also play an important role in HHO methods. We say that the $(d-1)$-dimensional subset $F \subset \overline{\Omega}$ is a mesh face if F is a subset of an affine hyperplane, say H_F, such that the following holds: (i) either there are two distinct mesh cells $T_-, T_+ \in \mathcal{T}$ such that

$$F = \partial T_- \cap \partial T_+ \cap H_F, \tag{1.4}$$

and F is called a (mesh) interface; (ii) or there is one mesh cell $T_- \in \mathcal{T}$ such that

$$F = \partial T_- \cap \partial \Omega \cap H_F, \tag{1.5}$$

and F is called a (mesh) boundary face. The interfaces are collected in the set $\mathcal{F}^\circ$, the boundary faces in the set $\mathcal{F}^\partial$, so that the set

$$\mathcal{F} := \mathcal{F}^\circ \cup \mathcal{F}^\partial \tag{1.6}$$

collects all the mesh faces. For a mesh cell $T \in \mathcal{T}$, $\mathcal{F}_T$ denotes the collection of the mesh faces composing its boundary ∂T. Notice that the above definition of the mesh faces implies that each mesh face is straight in 2D and planar in 3D. Hence, for every mesh cell $T \in \mathcal{T}$, $\boldsymbol{n}_{T|F}$ is a constant vector on every face $F \in \mathcal{F}_T$. Notice also that the definitions (1.4) and (1.5) do not allow for the case of several coplanar faces that could be shared by two cells or a cell and the boundary, respectively; this choice is only made for simplicity.

1.2.2 Discrete Unknowns

The discrete unknowns in HHO methods are polynomials attached to the mesh cells and to the mesh faces. The idea is that the cell polynomials approximate the exact solution in the mesh cells, and that the face polynomials approximate the trace of the exact solution on the mesh faces (although they are not the trace of the cell polynomials). To ease the exposition, we consider here the equal-order HHO method where the cell and face polynomials have the same degree. Variants are considered in Sect. 3.2.1.

Let $k \geq 0$ be the polynomial degree. Let $\mathbb{P}_d^k$ be the space composed of d-variate (real-valued) polynomials of total degree at most k. For every mesh cell $T \in \mathcal{T}$, $\mathbb{P}_d^k(T)$ denotes the space composed of the restriction to T of the polynomials in $\mathbb{P}_d^k$. To define the $(d-1)$-variate polynomial space attached to a mesh face $F \in \mathcal{F}$ (which is a subset of $\mathbb{R}^d$), we consider an affine geometric mapping $\boldsymbol{T}_F : \mathbb{R}^{d-1} \to H_F$ (recall that H_F is the affine hyperplane in $\mathbb{R}^d$ supporting F). Then we set

$$\mathbb{P}_{d-1}^k(F) := \mathbb{P}_{d-1}^k \circ (\boldsymbol{T}_F^{-1})_{|F}. \tag{1.7}$$

It is easy to see that the definition of $\mathbb{P}_{d-1}^k(F)$ is independent of the choice of the affine geometric mapping $\boldsymbol{T}_F$. (Notice that defining polynomials on the mesh faces is meaningful since we are assuming $d \geq 2$.)

Let us first consider a local viewpoint. For every mesh cell $T \in \mathcal{T}$, we set

$$\hat{V}_T^k := \mathbb{P}_d^k(T) \times \mathbb{P}_{d-1}^k(\mathcal{F}_T), \qquad \mathbb{P}_{d-1}^k(\mathcal{F}_T) := \bigtimes_{F \in \mathcal{F}_T} \mathbb{P}_{d-1}^k(F). \tag{1.8}$$

A generic element in $\hat{V}_T^k$ is denoted by $\hat{v}_T := (v_T, v_{\partial T})$. We shall systematically employ the hat notation to indicate a pair of (piecewise) functions, one attached to the mesh cell(s) and one to the mesh face(s). Notice that the trace of v_T on ∂T differs from $v_{\partial T}$; in particular, the former is a smooth function over ∂T, whereas the latter generally exhibits jumps from one face in $\mathcal{F}_T$ to an adjacent one. To define the global discrete HHO unknowns, we follow a similar paradigm; see Fig. 1.2.

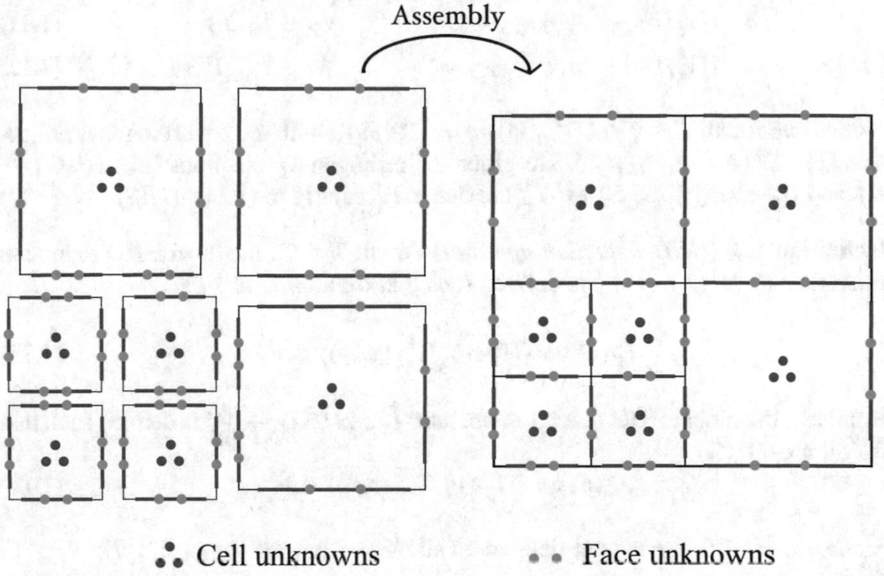

Fig. 1.2 Local (left) and global (right) unknowns for the HHO method ($d = 2, k = 1$). Each bullet on the faces and in the cells conventionally represents one basis function

Definition 1.1 (*HHO space*) The equal-order HHO space is defined as follows:

$$\hat{V}_h^k := V_{\mathcal{T}}^k \times V_{\mathcal{F}}^k, \qquad V_{\mathcal{T}}^k := \bigtimes_{T\in\mathcal{T}} \mathbb{P}_d^k(T), \qquad V_{\mathcal{F}}^k := \bigtimes_{F\in\mathcal{F}} \mathbb{P}_{d-1}^k(F). \tag{1.9}$$

We have $\dim(\hat{V}_h^k) = \binom{k+d}{d}\#\mathcal{T} + \binom{k+d-1}{d-1}\#\mathcal{F}$. □

A generic element in $\hat{V}_h^k$ is denoted by $\hat{v}_h := (v_{\mathcal{T}}, v_{\mathcal{F}})$ with $v_{\mathcal{T}} := (v_T)_{T\in\mathcal{T}}$ and $v_{\mathcal{F}} := (v_F)_{F\in\mathcal{F}}$. Notice that in general $v_{\mathcal{T}}$ is only piecewise smooth, i.e., it can jump across the mesh interfaces, and similarly $v_{\mathcal{F}}$ can jump from one mesh face to an adjacent one. Moreover, for all $\hat{v}_h \in \hat{V}_h^k$ and all $T \in \mathcal{T}$, it is convenient to localize the components of $\hat{v}_h$ associated with T and its faces by using the notation

$$\hat{v}_T := \big(v_T, v_{\partial T} := (v_F)_{F\in\mathcal{F}_T}\big) \in \hat{V}_T^k. \tag{1.10}$$

At this stage, a natural question that arises is how to reduce a generic function $v \in H^1(\Omega)$ (think of the weak solution to (1.2)) to some member of the discrete space $\hat{V}_h^k$. In the context of finite elements, this task is usually realized by means of the interpolation operator associated with the finite element. In the context of HHO methods, this task is realized in a simple way by considering L^2-orthogonal projections. Let $T \in \mathcal{T}$. Let $\Pi_T^k : L^2(T) \to \mathbb{P}_d^k(T)$ and $\Pi_{\partial T}^k : L^2(\partial T) \to \mathbb{P}_{d-1}^k(\mathcal{F}_T)$ be the L^2-orthogonal projections defined such that for all $v \in L^2(T)$ and all $w \in L^2(\partial T)$,

$$(\Pi_T^k(v) - v, p)_{L^2(T)} = 0 \qquad \forall p \in \mathbb{P}_d^k(T), \tag{1.11}$$

$$(\Pi_{\partial T}^k(w) - w, q)_{L^2(\partial T)} = 0 \qquad \forall q \in \mathbb{P}_{d-1}^k(\mathcal{F}_T). \tag{1.12}$$

Notice that for all $F \in \mathcal{F}_T$, $(\Pi_{\partial T}^k(w))_{|F} = \Pi_F^k(w_{|F})$, with the L^2-orthogonal projection $\Pi_F^k : L^2(F) \to \mathbb{P}_{d-1}^k(F)$. The global L^2-orthogonal projections $\Pi_{\mathcal{T}}^k : L^2(\Omega) \to V_{\mathcal{T}}^k$ and $\Pi_{\mathcal{F}}^k : L^2(\bigcup_{F\in\mathcal{F}} F) \to V_{\mathcal{F}}^k$ are defined similarly to (1.11)–(1.12).

Definition 1.2 (*HHO reduction operator*) For all $T \in \mathcal{T}$, the local HHO reduction operator $\hat{I}_T^k : H^1(T) \to \hat{V}_T^k$ is defined such that for all $v \in H^1(T)$,

$$\hat{I}_T^k(v) := (\Pi_T^k(v), \Pi_{\partial T}^k(v_{|\partial T})) \in \hat{V}_T^k. \tag{1.13}$$

Similarly, the global HHO reduction operator $\hat{I}_h^k : H^1(\Omega) \to \hat{V}_h^k$ is defined such that for all $v \in H^1(\Omega)$,

$$\hat{I}_h^k(v) := (\Pi_{\mathcal{T}}^k(v), \Pi_{\mathcal{F}}^k(v_{|\mathcal{F}})) \in \hat{V}_h^k. \tag{1.14}$$

Since $v \in H^1(\Omega)$, $v_{|\mathcal{F}}$ is well-defined on all the mesh faces composing $\mathcal{F}$. □

1.3 Local Reconstruction and Stabilization

Local reconstruction and stabilization operators associated with each mesh cell lie at the heart of HHO methods. The goal of this section is to present these two operators and their main properties. In the whole section, $T \in \mathcal{T}$ denotes a generic mesh cell.

1.3.1 Local Reconstruction

The main purpose of the reconstruction operator is to compute a gradient in the mesh cell $T \in \mathcal{T}$ given a pair of discrete unknowns $\hat{v}_T := (v_T, v_{\partial T}) \in \hat{V}_T^k$. Obviously, a simple possibility is to take the gradient of the cell unknown. However, as we shall now see, taking also into account the face unknowns leads to a reconstruction operator with better approximation properties.

To stay simple, we consider for the time being a local reconstruction operator $R_T : \hat{V}_T^k \to \mathbb{P}_d^{k+1}(T)$, so that the gradient is reconstructed locally as $\nabla R_T(\hat{v}_T) \in \nabla\mathbb{P}_d^{k+1}(T) \subset \mathbb{P}_d^k(T; \mathbb{R}^d)$ (see Sect. 3.1 for some variants).

Definition 1.3 (*Reconstruction*) The local reconstruction operator $R_T : \hat{V}_T^k \to \mathbb{P}_d^{k+1}(T)$ is such that for all $\hat{v}_T := (v_T, v_{\partial T}) \in \hat{V}_T^k$, the function $R_T(\hat{v}_T) \in \mathbb{P}_d^{k+1}(T)$ is uniquely defined by the following equations:

$$(\nabla R_T(\hat{v}_T), \nabla q)_{L^2(T)} = -(v_T, \Delta q)_{L^2(T)} + (v_{\partial T}, \boldsymbol{n}_T\cdot\nabla q)_{L^2(\partial T)}, \tag{1.15}$$
$$(R_T(\hat{v}_T), 1)_{L^2(T)} = (v_T, 1)_{L^2(T)}, \tag{1.16}$$

where (1.15) holds for all $q \in \mathbb{P}_d^{k+1}(T)^\perp := \{q \in \mathbb{P}_d^{k+1}(T) \mid (q, 1)_{L^2(T)} = 0\}$. □

Integrating by parts in (1.15) readily yields for all $q \in \mathbb{P}_d^{k+1}(T)^\perp$,

$$(\nabla R_T(\hat{v}_T), \nabla q)_{L^2(T)} = (\nabla v_T, \nabla q)_{L^2(T)} - (v_T - v_{\partial T}, \boldsymbol{n}_T\cdot\nabla q)_{L^2(\partial T)}. \tag{1.17}$$

Moreover, we notice that $R_T(\hat{v}_T) = v_T$ if $v_{\partial T} = v_{T|\partial T}$, i.e., $R_T(\hat{v}_T)$ is in $\mathbb{P}_d^{k+1}(T)$ and not just in $\mathbb{P}_d^k(T)$ only if $v_{\partial T} \neq v_{T|\partial T}$. In practice, computing $R_T(\hat{v}_T)$ requires choosing a basis of $\mathbb{P}_d^{k+1}(T)^\perp$, inverting the corresponding local stiffness matrix of size $\binom{k+d+1}{d} - 1$, and adjusting the mean-value of $R_T(\hat{v}_T)$ in T using (1.16).

To motivate the above definition of R_T, we show that the composed operator $R_T \circ \hat{I}_T^k$ enjoys a higher-order approximation property.

Lemma 1.4 (Elliptic projection) *We have $\mathcal{E}_T^{k+1} = R_T \circ \hat{I}_T^k$ where $\mathcal{E}_T^{k+1} : H^1(T) \to \mathbb{P}_d^{k+1}(T)$ is the elliptic projection uniquely defined such that for all $v \subset H^1(T)$,*

$$(\nabla \mathcal{E}_T^{k+1}(v), \nabla q)_{L^2(T)} = (\nabla v, \nabla q)_{L^2(T)}, \quad \forall q \in \mathbb{P}_d^{k+1}(T)^\perp, \tag{1.18}$$
$$(\mathcal{E}_T^{k+1}(v), 1)_{L^2(T)} = (v, 1)_{L^2(T)}. \tag{1.19}$$

Proof Consider an arbitrary function $v \in H^1(T)$ and to alleviate the notation, let us set $\phi := R_T(\hat{I}_T^k(v)) = R_T(\Pi_T^k(v), \Pi_{\partial T}^k(v_{|\partial T}))$. Using the definition (1.15) of R_T, we infer that

$$\begin{aligned}(\nabla\phi, \nabla q)_{L^2(T)} &= -(\Pi_T^k(v), \Delta q)_{L^2(T)} + (\Pi_{\partial T}^k(v_{|\partial T}), \boldsymbol{n}_T\cdot\nabla q)_{L^2(\partial T)} \\ &= -(v, \Delta q)_{L^2(T)} + (v, \boldsymbol{n}_T\cdot\nabla q)_{L^2(\partial T)} = (\nabla v, \nabla q)_{L^2(T)},\end{aligned}$$

for all $q \in \mathbb{P}_d^{k+1}(T)^\perp$, since $\Delta q \in \mathbb{P}_d^{k-1}(T) \subset \mathbb{P}_d^k(T)$ and $\boldsymbol{n}_T\cdot\nabla q \in \mathbb{P}_{d-1}^k(\mathcal{F}_T)$ (here, we use that all the faces are planar so that $\boldsymbol{n}_T$ is piecewise constant; projectors were removed owing to (1.11) and (1.12)). Moreover, we have

$$(\phi, 1)_{L^2(T)} = (R_T(\hat{I}_T^k(v)), 1)_{L^2(T)} = (\Pi_T^k(v), 1)_{L^2(T)} = (v, 1)_{L^2(T)},$$

owing to the definition of R_T and $\hat{I}_T^k$. These two identities prove that ϕ satisfies (1.18)–(1.19), so that $\phi = \mathcal{E}_T^{k+1}(v)$ for all $v \in H^1(T)$. Hence, $\mathcal{E}_T^{k+1} = R_T \circ \hat{I}_T^k$. □

1.3.2 Local Stabilization

The main issue with the reconstruction operator is that $\nabla R_T(\hat{v}_T) = \mathbf{0}$ does not imply that v_T and $v_{\partial T}$ are constant functions taking the same value. Indeed, since $\#\mathcal{F}_T \geq$

$d+1$, we have $\dim(\hat{V}_T^k) = \dim(\mathbb{P}_d^k) + \dim(\mathbb{P}_{d-1}^k)\#\mathcal{F}_T \geq \dim(\mathbb{P}_d^k) + \dim(\mathbb{P}_{d-1}^k)$ $(d+1)$, and the rank theorem together with Lemma 1.4 give $\dim(\ker(R_T)) = \dim(\hat{V}_T^k) - \dim(\mathrm{im}(R_T)) = \dim(\hat{V}_T^k) - \dim(\mathbb{P}_d^{k+1})$. Combining these two inequalities and since $\dim(\mathbb{P}_{d'}^l) = \binom{l+d'}{d'}$, this shows that $\dim(\ker(R_T)) \geq 1$.

To fix this issue, a local stabilization operator is introduced. Among various possibilities, we focus on an operator that maps $\hat{V}_T^k$ to face-based functions $S_{\partial T} : \hat{V}_T^k \to \mathbb{P}_{d-1}^k(\mathcal{F}_T)$ such that for all $\hat{v}_T \in \hat{V}_T^k$,

$$S_{\partial T}(\hat{v}_T) := \Pi_{\partial T}^k\Big(v_{T|\partial T} - v_{\partial T} + \big((I - \Pi_T^k)R_T(\hat{v}_T)\big)_{|\partial T}\Big), \tag{1.20}$$

where I is the identity operator. Letting $\delta_{\partial T} := v_{T|\partial T} - v_{\partial T}$ be the difference between the trace of the cell component and the face component on ∂T, we observe that

$$R_T(\hat{v}_T) = R_T(v_T, v_{T|\partial T}) - R_T(0, \delta_{\partial T}) = v_T - R_T(0, \delta_{\partial T}). \tag{1.21}$$

Since $v_T \in \mathbb{P}_d^k(T)$, the operator $S_{\partial T}$ in (1.20) can be rewritten as follows:

$$S_{\partial T}(\hat{v}_T) = \Pi_{\partial T}^k\Big(\delta_{\partial T} - \big((I - \Pi_T^k)R_T(0, \delta_{\partial T})\big)_{|\partial T}\Big). \tag{1.22}$$

This shows that $S_{\partial T}(\hat{v}_T)$ only depends (linearly) on the difference $(v_{T|\partial T} - v_{\partial T})$. The role of $S_{\partial T}$ is to help enforce the matching between the trace of the cell component and the face component. In the discrete problem, this matching is enforced in a least-squares manner (see Sect. 1.4.1). In practice, computing $S_{\partial T}(\hat{v}_T)$ requires to evaluate L^2-orthogonal projections in the cell and on its faces, which entails inverting the mass matrix in T, which is of size $\binom{k+d}{d}$, and inverting the mass matrix in each face $F \in \mathcal{F}_T$, which is of size $\binom{k+d-1}{d-1}$.

Let us finally state an important stability result motivating the introduction of the operator $S_{\partial T}$. To this purpose, we equip the space $\hat{V}_T^k$ with the following H^1-like seminorm: For all $\hat{v}_T \in \hat{V}_T^k$,

$$|\hat{v}_T|_{\hat{V}_T^k}^2 := \|\nabla v_T\|_{L^2(T)}^2 + h_T^{-1}\|v_T - v_{\partial T}\|_{L^2(\partial T)}^2. \tag{1.23}$$

Notice that $|\hat{v}_T|_{\hat{V}_T^k} = 0$ implies that v_T and $v_{\partial T}$ are constant functions taking the same value. Then, as shown in Sect. 2.2, there are $0 < \alpha \leq \omega < +\infty$, independent of the mesh size h, such that for all $T \in \mathcal{T}$ and all $\hat{v}_T \in \hat{V}_T^k$,

$$\alpha|\hat{v}_T|_{\hat{V}_T^k}^2 \leq \|\nabla R_T(\hat{v}_T)\|_{L^2(T)}^2 + h_T^{-1}\|S_{\partial T}(\hat{v}_T)\|_{L^2(\partial T)}^2 \leq \omega|\hat{v}_T|_{\hat{V}_T^k}^2. \tag{1.24}$$

1.3.3 Example: Lowest-Order Case

Let us briefly illustrate the above reconstruction and stabilization operators in the lowest-order case where $k = 0$. Then, for all $\hat{v}_T := (v_T, v_{\partial T}) \in \hat{V}_T^0$, v_T is constant on T and $v_{\partial T}$ is piecewise constant on ∂T. Moreover, $\nabla R_T(\hat{v}_T)$ is a constant vector in T, $R_T(\hat{v}_T) \in \mathbb{P}_d^1(T)$, and $S_{\partial T}(\hat{v}_T)$ is piecewise constant on ∂T.

Proposition 1.5 (Lowest-order realization) *Assume $k = 0$. Let $T \in \mathcal{T}$. Let $\boldsymbol{x}_T$ be the barycenter of T and $\boldsymbol{x}_F$ that of the face $F \in \mathcal{F}_T$. For all $\hat{v}_T := (v_T, v_{\partial T}) \in \hat{V}_T^0$, setting $v_F := v_{\partial T|F}$ for all $F \in \mathcal{F}_T$, we have*

$$\nabla R_T(\hat{v}_T) = \sum_{F \in \mathcal{F}_T} \frac{|F|}{|T|}(v_F - v_T)\boldsymbol{n}_{T|F}, \tag{1.25}$$

$$R_T(\hat{v}_T)(\boldsymbol{x}) = v_T + \nabla R_T(\hat{v}_T)\cdot(\boldsymbol{x} - \boldsymbol{x}_T), \qquad \forall \boldsymbol{x} \in T, \tag{1.26}$$

$$S_{\partial T}(\hat{v}_T)_{|F} = v_T - v_F - \nabla R_T(\hat{v}_T)\cdot(\boldsymbol{x}_T - \boldsymbol{x}_F), \qquad \forall F \in \mathcal{F}_T. \tag{1.27}$$

Proof The proof revolves around the fact that any polynomial $q \in \mathbb{P}_d^1(T)$ is such that $q(\boldsymbol{x}) = \bar{q}_T + \boldsymbol{G}_q\cdot(\boldsymbol{x} - \boldsymbol{x}_T)$ for all $\boldsymbol{x} \in T$, where $\bar{q}_T$ is the mean-value of q in T and $\boldsymbol{G}_q := \nabla q$ is a constant vector in T. Using (1.17) and $\nabla v_T = \boldsymbol{0}$ gives for all $q \in \mathbb{P}_d^1(T)$,

$$\begin{aligned}
|T|\nabla R_T(\hat{v}_T)\cdot\boldsymbol{G}_q &= (\nabla R_T(\hat{v}_T), \nabla q)_{L^2(T)} \\
&= (\nabla v_T, \nabla q)_{L^2(T)} - (v_T - v_{\partial T}, \boldsymbol{n}_T\cdot\nabla q)_{L^2(\partial T)} \\
&= \sum_{F \in \mathcal{F}_T} |F|(v_F - v_T)\boldsymbol{n}_{T|F}\cdot\boldsymbol{G}_q.
\end{aligned}$$

Since $\boldsymbol{G}_q$ can be chosen arbitrarily in $\mathbb{R}^d$, this proves (1.25). The expression (1.26) then follows from the above characterization of polynomials in $\mathbb{P}_d^1(T)$ and (1.16). Finally, since $\Pi_T^0(R_T(\hat{v}_T)) = v_T$ and $\Pi_F^0(R_T(\hat{v}_T)) = v_T + \nabla R_T(\hat{v}_T)\cdot(\boldsymbol{x}_F - \boldsymbol{x}_T)$ for all $F \in \mathcal{F}_T$, inserting these expressions into (1.20) yields (1.27). □

The idea of reconstructing a gradient in each mesh cell by means of (1.25) and adding a stabilization proportional to (1.27) has been considered in the hybrid finite volume (HFV) method from [97].

1.4 Assembly and Static Condensation

In this section, we present the discrete problem resulting from the HHO approximation of the weak problem (1.2). We then highlight the algebraic realization of the discrete problem and show that the cell unknowns can be eliminated locally by a Schur complement technique often called static condensation.

1.4.1 The Discrete Problem

The discrete problem is formulated by means of a discrete bilinear form $a_h : \hat{V}_h^k \times \hat{V}_h^k \to \mathbb{R}$ which is assembled cellwise in the same spirit as in the finite element method. Thus, for all $\hat{v}_h, \hat{w}_h \in \hat{V}_h^k$, we set

$$a_h(\hat{v}_h, \hat{w}_h) := \sum_{T\in\mathcal{T}} a_T(\hat{v}_T, \hat{w}_T), \tag{1.28}$$

where we recall that $\hat{v}_T \in \hat{V}_T^k$ (resp., $\hat{w}_T \in \hat{V}_T^k$) collects the components of $\hat{v}_h$ (resp., $\hat{w}_h$) associated with the cell $T \in \mathcal{T}$ and the faces $F \in \mathcal{F}_T$ composing its boundary. The local bilinear form $a_T : \hat{V}_T^k \times \hat{V}_T^k \to \mathbb{R}$ is devised by using the local reconstruction and stabilization operators introduced in the previous section by setting

$$a_T(\hat{v}_T, \hat{w}_T) := (\nabla R_T(\hat{v}_T), \nabla R_T(\hat{w}_T))_{L^2(T)} + h_T^{-1}(S_{\partial T}(\hat{v}_T), S_{\partial T}(\hat{w}_T))_{L^2(\partial T)}. \tag{1.29}$$

The first term on the right-hand side is the counterpart of the local term $(\nabla v, \nabla w)_{L^2(T)}$ in the exact bilinear form a, whereas the second term acts as a stabilization that weakly enforces the matching between the trace of the cell unknowns and the face unknowns. Notice that the scaling by h_T^{-1} makes both terms in (1.29) dimensionally consistent and, at the same time, ensures optimally-decaying error estimates (see Chap. 2). Defining the piecewise polynomial space $\mathbb{P}_d^{k+1}(\mathcal{T}) := \{v \in L^2(\Omega) \mid v_{|T} \in \mathbb{P}_d^{k+1}(T),\ \forall T \in \mathcal{T}\}$, the global reconstruction operator $R_{\mathcal{T}} : \hat{V}_h^k \to \mathbb{P}_d^{k+1}(\mathcal{T})$ is such that

$$R_{\mathcal{T}}(\hat{v}_h)_{|T} := R_T(\hat{v}_T), \quad \forall \hat{v}_h \in \hat{V}_h^k, \quad \forall T \in \mathcal{T}. \tag{1.30}$$

We also define the global stabilization bilinear form $s_h : \hat{V}_h^k \times \hat{V}_h^k \to \mathbb{R}$ such that

$$s_h(\hat{v}_h, \hat{w}_h) := \sum_{T\in\mathcal{T}} h_T^{-1}(S_{\partial T}(\hat{v}_T), S_{\partial T}(\hat{w}_T))_{L^2(\partial T)}. \tag{1.31}$$

The discrete bilinear form a_h can then be rewritten as follows:

$$a_h(\hat{v}_h, \hat{w}_h) = (\nabla_{\mathcal{T}} R_{\mathcal{T}}(\hat{v}_h), \nabla_{\mathcal{T}} R_{\mathcal{T}}(\hat{w}_h))_{L^2(\Omega)} + s_h(\hat{v}_h, \hat{w}_h), \tag{1.32}$$

with the broken gradient operator $\nabla_{\mathcal{T}}$ acting locally in every mesh cell.

We enforce strongly the homogeneous Dirichlet boundary condition by zeroing out the discrete unknowns associated with the boundary faces, i.e., we consider the subspace

$$\hat{V}_{h,0}^k := V_{\mathcal{T}}^k \times V_{\mathcal{F},0}^k, \quad V_{\mathcal{F},0}^k := \{v_{\mathcal{F}} \in V_{\mathcal{F}}^k \mid v_F = 0,\ \forall F \in \mathcal{F}^\partial\}. \tag{1.33}$$

The discrete problem is as follows:

$$\begin{cases} \text{Find } \hat{u}_h \in \hat{V}^k_{h,0} \text{ such that} \\ a_h(\hat{u}_h, \hat{w}_h) = \ell(w_{\mathcal{T}}), \quad \forall \hat{w}_h := (w_{\mathcal{T}}, w_{\mathcal{F}}) \in \hat{V}^k_{h,0}. \end{cases} \tag{1.34}$$

Notice that only the cell component of the test function $\hat{w}_h$ is used to evaluate the load term since $\ell(w_{\mathcal{T}}) := (f, w_{\mathcal{T}})_{L^2(\Omega)} = \sum_{T\in\mathcal{T}} (f, w_T)_{L^2(T)}$ (we keep the same symbol ℓ for simplicity). A more subtle treatment of the load term is needed if one works with loads in the dual Sobolev space $H^{-1}(\Omega)$ (see [95] for further insight).

To establish the well-posedness of (1.34), we prove that the bilinear form a_h is coercive on $\hat{V}^k_{h,0}$. To this purpose, we equip this space with a suitable norm. Recall the H^1-like seminorm $|\cdot|_{\hat{V}^k_T}$ defined in (1.23).

Lemma 1.6 (Norm) *The following map defines a norm on* $\hat{V}^k_{h,0}$:

$$\hat{V}^k_{h,0} \ni \hat{v}_h \longmapsto \|\hat{v}_h\|_{\hat{V}^k_{h,0}} := \left(\sum_{T\in\mathcal{T}} |\hat{v}_T|^2_{\hat{V}^k_T}\right)^{\frac{1}{2}} \in [0, +\infty). \tag{1.35}$$

Proof The only nontrivial property to verify is the definiteness of the map. Let $\hat{v}_h \in \hat{V}^k_{h,0}$ be such that $\|\hat{v}_h\|_{\hat{V}^k_{h,0}} = 0$, i.e., $|\hat{v}_T|_{\hat{V}^k_T} = 0$ for all $T \in \mathcal{T}$. Owing to (1.23), we infer that v_T and $v_{\partial T}$ are constant functions taking the same value in each mesh cell. On cells having a boundary face, this value must be zero since $v_{\mathcal{F}}$ vanishes on the boundary faces. We can repeat the argument for the cells sharing an interface with those cells, and we can move inward and reach all the cells in $\mathcal{T}$ by repeating this process a finite number of times. Thus, $\hat{v}_h = (0, 0) \in \hat{V}^k_{h,0}$. □

Lemma 1.7 (Coercivity and well-posedness) *The bilinear form* a_h *is coercive on* $\hat{V}^k_{h,0}$, *and the discrete problem (1.34) is well-posed.*

Proof The coercivity of a_h follows by summing the lower bound in (1.24) over the mesh cells, which yields

$$a_h(\hat{v}_h, \hat{v}_h) \geq \alpha \|\hat{v}_h\|^2_{\hat{V}^k_{h,0}}, \quad \forall \hat{v}_h \in \hat{V}^k_{h,0}. \tag{1.36}$$

Well-posedness is a consequence of the Lax–Milgram lemma. □

Standard convexity arguments show that the weak solution $u \in V = H^1_0(\Omega)$ to (1.2) is the unique minimizer in V of the energy functional

$$\mathfrak{E} : V \ni v \longmapsto \frac{1}{2}\|\nabla v\|^2_{\boldsymbol{L}^2(\Omega)} - \ell(v) \in \mathbb{R}. \tag{1.37}$$

The HHO solution $\hat{u}_h \in \hat{V}^k_{h,0}$ of (1.34) can also be characterized as the unique minimizer in $\hat{V}^k_{h,0}$ of a suitable energy functional, namely

$$\mathfrak{E}_h : \hat{V}^k_{h,0} \ni \hat{v}_h \longmapsto \frac{1}{2}\|\nabla_{\mathcal{T}} R_{\mathcal{T}}(\hat{v}_h)\|^2_{L^2(\Omega)} + \frac{1}{2}s_h(\hat{v}_h, \hat{v}_h) - \ell(v_{\mathcal{T}}) \in \mathbb{R}. \tag{1.38}$$

Proposition 1.8 (HHO energy minimization) *Let $\mathfrak{E}_h : \hat{V}^k_{h,0} \to \mathbb{R}$ be defined in (1.38). Then $\hat{u}_h \in \hat{V}^k_{h,0}$ solves (1.34) if and only if $\hat{u}_h$ minimizes $\mathfrak{E}_h$ in $\hat{V}^k_{h,0}$.*

Proof Owing to the coercivity of the discrete bilinear form a_h established in Lemma 1.7, the discrete energy functional $\mathfrak{E}_h$ is strongly convex in $\hat{V}^k_{h,0}$. Moreover, this functional is Fréchet-differentiable at any $\hat{v}_h \in \hat{V}^k_{h,0}$, and a straightforward calculation shows that for all $\hat{w}_h \in \hat{V}^k_{h,0}$, $D\mathfrak{E}_h(\hat{v}_h)[\hat{w}_h] = a_h(\hat{v}_h, \hat{w}_h) - \ell(w_{\mathcal{T}})$. This proves the claimed equivalence. □

To streamline the presentation, we postpone the statement and proof of the main error estimates regarding the HHO method to the next chapter. At this stage, we merely announce that, under reasonable assumptions, the (broken) H^1-seminorm of the error decays as $O(h^{k+1})$ and the L^2-norm of the error decays as $O(h^{k+2})$ where h denotes the mesh size. More precise statements can be found in Sect. 2.4–2.5. A residual-based a posteriori error analysis can be found in [83].

Remark 1.9 (*Face unknowns*) Consider the energy functional $\mathfrak{E}_{\mathcal{F},0}(v_{\mathcal{T}}, \cdot) : V^k_{\mathcal{F},0} \to \mathbb{R}$ such that $\mathfrak{E}_{\mathcal{F},0}(v_{\mathcal{T}}, \cdot) = \frac{1}{2}\|\nabla R_{\mathcal{T}}(v_{\mathcal{T}}, \cdot)\|^2_{L^2(\Omega)} + \frac{1}{2}s_h((v_{\mathcal{T}}, \cdot), (v_{\mathcal{T}}, \cdot))$ for all $v_{\mathcal{T}} \in V^k_{\mathcal{T}}$. Elementary arguments show that $\mathfrak{E}_{\mathcal{F},0}(v_{\mathcal{T}}, \cdot)$ admits a unique minimizer in $V^k_{\mathcal{F},0}$ which we denote $v^*_{\mathcal{F}}(v_{\mathcal{T}}) \in V^k_{\mathcal{F},0}$ for all $v_{\mathcal{T}} \in V^k_{\mathcal{T}}$. Let $\mathfrak{E}_{\mathcal{T}} : V^k_{\mathcal{T}} \to \mathbb{R}$ be the energy functional such that $\mathfrak{E}_{\mathcal{T}}(v_{\mathcal{T}}) := \mathfrak{E}_h(v_{\mathcal{T}}, v^*_{\mathcal{F}}(v_{\mathcal{T}}))$. Then, $\hat{u}_h = (u_{\mathcal{T}}, u_{\mathcal{F}}) \in \hat{V}^k_{h,0}$ solves (1.34) if and only if $u_{\mathcal{F}} = v^*_{\mathcal{F}}(u_{\mathcal{T}})$ and $u_{\mathcal{T}}$ is the unique minimizer of $\mathfrak{E}_{\mathcal{T}}$ in $V^k_{\mathcal{T}}$. □

1.4.2 Algebraic Realization

Let $N^k_{\mathcal{T}} := \dim(V^k_{\mathcal{T}}) = \binom{k+d}{d}\#\mathcal{T}$ and let $N^k_{\mathcal{F},0} := \dim(V^k_{\mathcal{F},0}) = \binom{k+d-1}{d-1}\#\mathcal{F}^\circ$. Let $(\mathsf{U}_{\mathcal{T}}, \mathsf{U}_{\mathcal{F}}) \in \mathbb{R}^{N^k_{\mathcal{T}} \times N^k_{\mathcal{F},0}}$ be the component vectors of the discrete solution $\hat{u}_h := (u_{\mathcal{T}}, u_{\mathcal{F}}) \in \hat{V}^k_{h,0}$ once bases $\{\varphi_i\}_{1\le i\le N^k_{\mathcal{T}}}$ and $\{\psi_j\}_{1\le j\le N^k_{\mathcal{F},0}}$ for $V^k_{\mathcal{T}}$ and $V^k_{\mathcal{F},0}$, respectively, have been chosen. (Notice that the components of $\mathsf{U}_{\mathcal{F}}$ are attached only to the mesh interfaces.) Let $\mathsf{F}_{\mathcal{T}} \in \mathbb{R}^{N^k_{\mathcal{T}}}$ have components given by $\mathsf{F}_i := (f, \varphi_i)_{L^2(\Omega)}$ for all $1 \le i \le N^k_{\mathcal{T}}$. The algebraic realization of (1.34) is

$$\begin{bmatrix} \mathsf{A}_{\mathcal{T}\mathcal{T}} & \mathsf{A}_{\mathcal{T}\mathcal{F}} \\ \mathsf{A}_{\mathcal{F}\mathcal{T}} & \mathsf{A}_{\mathcal{F}\mathcal{F}} \end{bmatrix} \begin{bmatrix} \mathsf{U}_{\mathcal{T}} \\ \mathsf{U}_{\mathcal{F}} \end{bmatrix} = \begin{bmatrix} \mathsf{F}_{\mathcal{T}} \\ 0 \end{bmatrix}, \tag{1.39}$$

where the symmetric positive-definite stiffness matrix A is of size $N^k_{\mathcal{T}} + N^k_{\mathcal{F},0}$ and is composed of the blocks $\mathsf{A}_{\mathcal{T}\mathcal{T}}, \mathsf{A}_{\mathcal{T}\mathcal{F}}, \mathsf{A}_{\mathcal{F}\mathcal{T}}, \mathsf{A}_{\mathcal{F}\mathcal{F}}$ associated with the bilinear form a_h and the cell and face basis functions. Assume that the basis functions associated with a given cell or face are ordered consecutively. Then the submatrix $\mathsf{A}_{\mathcal{T}\mathcal{T}}$ is block-diagonal, whereas this is not the case for the submatrix $\mathsf{A}_{\mathcal{F}\mathcal{F}}$ since the entries attached to faces belonging to the same cell are coupled together. A computationally effective way to solve the linear system (1.39) is to eliminate locally the cell unknowns and solve first for the face unknowns. Defining the Schur complement matrix

$$\mathsf{A}^{\mathrm{S}}_{\mathcal{F}\mathcal{F}} := \mathsf{A}_{\mathcal{F}\mathcal{F}} - \mathsf{A}_{\mathcal{F}\mathcal{T}}\mathsf{A}^{-1}_{\mathcal{T}\mathcal{T}}\mathsf{A}_{\mathcal{T}\mathcal{F}}, \tag{1.40}$$

the global transmission problem coupling all the face unknowns is

$$\mathsf{A}^{\mathrm{S}}_{\mathcal{F}\mathcal{F}}\mathsf{U}_{\mathcal{F}} = -\mathsf{A}_{\mathcal{F}\mathcal{T}}\mathsf{A}^{-1}_{\mathcal{T}\mathcal{T}}\mathsf{F}_{\mathcal{T}}. \tag{1.41}$$

This linear system is only of size $N^k_{\mathcal{F},0}$. Once it is solved, one recovers locally the cell unknowns by using that $\mathsf{U}_{\mathcal{T}} = \mathsf{A}^{-1}_{\mathcal{T}\mathcal{T}}(\mathsf{F}_{\mathcal{T}} - \mathsf{A}_{\mathcal{T}\mathcal{F}}\mathsf{U}_{\mathcal{F}})$. This procedure is called static condensation.

It can be instructive to reformulate the above manipulations by working directly on the discrete bilinear forms a_T and the discrete HHO unknowns. To this purpose, for every mesh cell $T \in \mathcal{T}$, we define $U_\mu \in \mathbb{P}^k_d(T)$ for all $\mu \in \mathbb{P}^k_{d-1}(\mathcal{F}_T)$, and we define $U_r \in \mathbb{P}^k_d(T)$ for all $r \in L^2(T)$ as follows:

$$a_T((U_\mu, 0), (q, 0)) := -a_T((0, \mu), (q, 0)), \quad \forall q \in \mathbb{P}^k_d(T), \tag{1.42}$$

$$a_T((U_r, 0), (q, 0)) := (r, q)_{L^2(T)}, \quad \forall q \in \mathbb{P}^k_d(T). \tag{1.43}$$

These problems are well-posed since a_T is coercive on $\mathbb{P}^k_d(T) \times \{0\}$ owing to (1.24).

Proposition 1.10 (Transmission problem) *The pair $\hat{u}_h := (u_{\mathcal{T}}, u_{\mathcal{F}}) \in \hat{V}^k_{h,0}$ solves the HHO problem (1.34) if and only if the cell component satisfies $u_T = U_{u_{\partial T}} + U_{f_{|T}}$ for all $T \in \mathcal{T}$, and the face component $u_{\mathcal{F}} \in V^k_{\mathcal{F},0}$ solves the following global transmission problem:*

$$\sum_{T\in\mathcal{T}} a_T((U_{u_{\partial T}}, u_{\partial T}), (U_{w_{\partial T}}, w_{\partial T})) = \sum_{T\in\mathcal{T}} (f, U_{w_{\partial T}})_{L^2(T)}, \quad \forall w_{\mathcal{F}} \in V^k_{\mathcal{F},0}. \tag{1.44}$$

Proof (i) Assume that $\hat{u}_h$ solves (1.34). Let $T \in \mathcal{T}$ and $w_T \in \mathbb{P}^k_d(T)$. Since $a_T((u_T, u_{\partial T}), (w_T, 0)) = (f, w_T)_{L^2(T)} = a_T((U_{f_{|T}}, 0), (w_T, 0))$, we infer that

$$a_T((u_T - U_{f_{|T}}, u_{\partial T}), (w_T, 0)) = 0 = a_T((U_{u_{\partial T}}, u_{\partial T}), (w_T, 0)),$$

showing that $u_T - U_{f_{|T}} = U_{u_{\partial T}}$. This implies that for all $w_{\partial T} \in \mathbb{P}^k_{d-1}(\mathcal{F}_T)$,

$$\begin{aligned}
&a_T((U_{u_{\partial T}}, u_{\partial T}), (U_{w_{\partial T}}, w_{\partial T})) \\
&= a_T((u_T, u_{\partial T}), (U_{w_{\partial T}}, w_{\partial T})) - a_T((U_{f_{|T}}, 0), (U_{w_{\partial T}}, w_{\partial T})) \\
&= a_T((u_T, u_{\partial T}), (U_{w_{\partial T}}, w_{\partial T})),
\end{aligned}$$

where we used the symmetry of a_T and $a_T(U_{w_{\partial T}}, w_{\partial T}), (q,0)) = 0$ for all $q \in \mathbb{P}_d^k(T)$. Summing over $T \in \mathcal{T}$ and using (1.34) shows that $u_{\mathcal{F}}$ solves (1.44).
(ii) Assume that $u_{\mathcal{F}}$ solves (1.44). Let $\hat{w}_h \in \hat{V}_{h,0}^k$. Setting $\hat{u}_T := (u_T, u_{\partial T}) := (U_{f_{|T}}, 0) + (U_{u_{\partial T}}, u_{\partial T})$ for all $T \in \mathcal{T}$, we infer that

$$
\begin{aligned}
a_T(\hat{u}_T, \hat{w}_T) &= a_T((U_{f_{|T}}, 0) + (U_{u_{\partial T}}, u_{\partial T}), (w_T - U_{w_{\partial T}}, 0)) \\
&\quad + a_T((U_{f_{|T}}, 0) + (U_{u_{\partial T}}, u_{\partial T}), (U_{w_{\partial T}}, w_{\partial T})) \\
&= a_T((U_{f_{|T}}, 0), (w_T - U_{w_{\partial T}}, 0)) + (f, U_{w_{\partial T}})_{L^2(T)} \\
&\quad + a_T((U_{u_{\partial T}}, u_{\partial T}), (U_{w_{\partial T}}, w_{\partial T})) - (f, U_{w_{\partial T}})_{L^2(T)} \\
&= (f, w_T)_{L^2(T)} + a_T((U_{u_{\partial T}}, u_{\partial T}), (U_{w_{\partial T}}, w_{\partial T})) - (f, U_{w_{\partial T}})_{L^2(T)},
\end{aligned}
$$

using that $a_T((U_{u_{\partial T}}, u_{\partial T}), (y_T, 0)) = 0$ for all $y_T \in \mathbb{P}_d^k(T)$, a similar argument for $(U_{w_{\partial T}}, w_{\partial T})$ together with the symmetry of a_T, and the definition of $U_{f_{|T}}$. Summing over $T \in \mathcal{T}$ and using (1.44) shows that $\hat{u}_h$ solves (1.34). □

1.5 Flux Recovery and Embedding into HDG Methods

In this section, following [62], we uncover equilibrated fluxes in the HHO method. These fluxes, which are associated with all the faces of every mesh cell, are in equilibrium at every mesh interface and are balanced in every mesh cell with the source term. With these fluxes in hand, we embed HHO methods into the broad class of hybridizable discontinuous Galerkin (HDG) methods.

1.5.1 Flux Recovery

Let $\tilde{S}_{\partial T} : \mathbb{P}_{d-1}^k(\mathcal{F}_T) \to \mathbb{P}_{d-1}^k(\mathcal{F}_T)$ for all $T \in \mathcal{T}$ be defined such that

$$
\tilde{S}_{\partial T}(\mu) := \Pi_{\partial T}^k\Big(\mu - \big((I - \Pi_T^k) R_T(0, \mu)\big)_{|\partial T}\Big), \tag{1.45}
$$

so that the stabilization operator satisfies $S_{\partial T}(\hat{v}_T) = \tilde{S}_{\partial T}(v_{T|\partial T} - v_{\partial T})$ (see (1.22)). By definition, the adjoint of $\tilde{S}_{\partial T}$, say $\tilde{S}_{\partial T}^* : \mathbb{P}_{d-1}^k(\mathcal{F}_T) \to \mathbb{P}_{d-1}^k(\mathcal{F}_T)$, is such that $(\tilde{S}_{\partial T}^*(\lambda), \mu)_{L^2(\partial T)} = (\lambda, \tilde{S}_{\partial T}(\mu))_{L^2(\partial T)}$ for all $\lambda, \mu \in \mathbb{P}_{d-1}^k(\mathcal{F}_T)$. The numerical fluxes of a pair $\hat{v}_h \in \hat{V}_h^k$ at the boundary of every mesh cell $T \in \mathcal{T}$ are defined as

$$
\phi_{\partial T}(\hat{v}_T) := -\boldsymbol{n}_T \cdot \nabla R_T(\hat{v}_T)_{|\partial T} + h_T^{-1}(\tilde{S}_{\partial T}^* \circ \tilde{S}_{\partial T})(v_{T|\partial T} - v_{\partial T}) \in \mathbb{P}_{d-1}^k(\mathcal{F}_T). \tag{1.46}
$$

Proposition 1.11 (HHO rewriting with fluxes) *Let $\hat{u}_h \in \hat{V}^k_{h,0}$ solve (1.34) and let the numerical fluxes $\phi_{\partial T}(\hat{u}_T) \in \mathbb{P}^k_{d-1}(\mathcal{F}_T)$ be defined as in (1.46) for all $T \in \mathcal{T}$. The following holds:*
(i) *Equilibrium at every mesh interface $F = \partial T_- \cap \partial T_+ \cap H_F \in \mathcal{F}^\circ$:*

$$\phi_{\partial T_-}(\hat{u}_{T_-})_{|F} + \phi_{\partial T_+}(\hat{u}_{T_+})_{|F} = 0. \tag{1.47}$$

(ii) *Balance with the source term in every mesh cell $T \in \mathcal{T}$:*

$$(\nabla R_T(\hat{u}_T), \nabla q)_{\boldsymbol{L}^2(T)} + (\phi_{\partial T}(\hat{u}_T), q)_{L^2(\partial T)} = (f, q)_{L^2(T)}, \quad \forall q \in \mathbb{P}^k_d(T). \tag{1.48}$$

(iii) *(1.47)–(1.48) are an equivalent rewriting of (1.34) that fully characterizes the HHO solution $\hat{u}_h \in \hat{V}^k_{h,0}$.*

Proof (i) Let $F \in \mathcal{F}^\circ$. The identity (1.47) is proved by taking a test function $\hat{w}_h$ in (1.34) whose only nonzero component is attached to the interface F. Let $w_F \in \mathbb{P}^k_{d-1}(F)$ and take $\hat{w}_h := (0, w_{\mathcal{F}})$ with $w_{\mathcal{F}} := (\delta_{F,F'} w_F)_{F' \in \mathcal{F}}$, where $\delta_{F,F'}$ is the Kronecker delta. This is a legitimate test function, i.e., $\hat{w}_h \in \hat{V}^k_{h,0}$. Letting $\mathcal{T}_F := \{T_-, T_+\}$, and using the definitions of a_h and a_T, we infer that

$$\begin{aligned} 0 &= \sum_{T \in \mathcal{T}_F} a_T(\hat{u}_T, (0, w_{\partial T})) \\ &= \sum_{T \in \mathcal{T}_F} (\nabla R_T(\hat{u}_T), \nabla R_T(0, w_{\partial T}))_{\boldsymbol{L}^2(T)} - h_T^{-1}(\tilde{S}_{\partial T}(u_{T|\partial T} - u_{\partial T}), \tilde{S}_{\partial T}(w_{\partial T}))_{L^2(\partial T)}. \end{aligned}$$

Using that $(\nabla R_T(\hat{u}_T), \nabla R_T(0, w_{\partial T}))_{\boldsymbol{L}^2(T)} = (\boldsymbol{n}_T{\cdot}\nabla R_T(\hat{u}_T), w_F)_{L^2(F)}$ and the definition of the adjoint operator $\tilde{S}^*_{\partial T}$ then gives

$$\begin{aligned} 0 &= \sum_{T \in \mathcal{T}_F} (\boldsymbol{n}_T{\cdot}\nabla R_T(\hat{u}_T), w_F)_{L^2(F)} - h_T^{-1}((\tilde{S}^*_{\partial T} \circ \tilde{S}_{\partial T})(u_{T|\partial T} - u_{\partial T}), w_F)_{L^2(F)} \\ &= \sum_{T \in \mathcal{T}_F} (\phi_{\partial T}(\hat{u}_T), w_F)_{L^2(F)}. \end{aligned}$$

Since $\phi_{\partial T}(\hat{u}_T)_{|F} \in \mathbb{P}^k_{d-1}(F)$ for all $T \in \mathcal{T}_F$ and w_F is arbitrary in $\mathbb{P}^k_{d-1}(F)$, we conclude that (1.47) holds true.
(ii) Let $T \in \mathcal{T}$. The identity (1.48) is proved by taking a test function $\hat{w}_h$ in (1.34) whose only nonzero component is attached to the mesh cell T. Let $q \in \mathbb{P}^k_d(T)$ and take $\hat{w}_h := (w_{\mathcal{T}}, 0)$ with $w_{\mathcal{T}} := (\delta_{T,T'} q)_{T' \in \mathcal{T}}$, so that $w_T = q$. This is a legitimate test function, i.e., $\hat{w}_h \in \hat{V}^k_{h,0}$. Since (1.15) implies that $(\nabla R_T(\hat{u}_T), \nabla R_T(\hat{w}_T))_{\boldsymbol{L}^2(T)} = (\nabla R_T(\hat{u}_T), \nabla q)_{\boldsymbol{L}^2(T)} - (\boldsymbol{n}_T{\cdot}\nabla R_T(\hat{u}_T), q)_{L^2(\partial T)}$, we have

$$
\begin{aligned}
(f,q)_{L^2(T)} &= a_T(\hat{u}_T,\hat{w}_T)\\
&= (\nabla R_T(\hat{u}_T),\nabla R_T(\hat{w}_T))_{L^2(T)} + h_T^{-1}(\tilde{S}_{\partial T}(u_{T|\partial T}-u_{\partial T}),\tilde{S}_{\partial T}(q))_{L^2(\partial T)}\\
&= (\nabla R_T(\hat{u}_T),\nabla q)_{L^2(T)} + (-\boldsymbol{n}_T{\cdot}\nabla R_T(\hat{u}_T) + h_T^{-1}(\tilde{S}^*_{\partial T}\circ\tilde{S}_{\partial T})(u_{T|\partial T}-u_{\partial T}),q)_{L^2(\partial T)}\\
&= (\nabla R_T(\hat{u}_T),\nabla q)_{L^2(T)} + (\phi_{\partial T}(\hat{u}_T),q)_{L^2(\partial T)}.
\end{aligned}
$$

(iii) The last assertion is a direct consequence of the above two proofs since the considered test functions span $\hat{V}^k_{h,0}$. □

1.5.2 Embedding into HDG Methods

HDG methods were introduced in [64] (see also [61] for an overview). In such methods, one approximates a triple, whereas one approximates a pair in HHO methods. Let us consider the dual variable $\boldsymbol{\sigma} := -\nabla u$ (sometimes called flux), the primal variable u, and its trace $\lambda := u_{|\mathcal{F}}$ on the mesh faces. HDG methods approximate the triple $(\boldsymbol{\sigma}, u, \lambda)$ by introducing some local spaces $\boldsymbol{S}_T$, V_T, and V_F for all $T \in \mathcal{T}$ and all $F \in \mathcal{F}$, and by defining a numerical flux trace that includes a stabilization operator. Defining the global spaces

$$\boldsymbol{S}_{\mathcal{T}} := \{\boldsymbol{\tau}_{\mathcal{T}} := (\boldsymbol{\tau}_T)_{T\in\mathcal{T}} \in \boldsymbol{L}^2(\Omega) \mid \boldsymbol{\tau}_T \in \boldsymbol{S}_T,\ \forall T \in \mathcal{T}\}, \tag{1.49}$$

$$V_{\mathcal{T}} := \{v_{\mathcal{T}} := (v_T)_{T\in\mathcal{T}} \in L^2(\Omega) \mid v_T \in V_T,\ \forall T \in \mathcal{T}\}, \tag{1.50}$$

$$V_{\mathcal{F}} := \{\mu_{\mathcal{F}} := (\mu_F)_{F\in\mathcal{F}} \in L^2(\mathcal{F}) \mid \mu_F \in V_F,\ \forall F \in \mathcal{F}\}, \tag{1.51}$$

as well as $V_{\mathcal{F},0} := \{\mu_{\mathcal{F}} \in V_{\mathcal{F}} \mid \mu_F = 0,\ \forall F \in \mathcal{F}^\partial\}$, the HDG method consists in seeking the triple $(\boldsymbol{\sigma}_{\mathcal{T}}, u_{\mathcal{T}}, \lambda_{\mathcal{F}}) \in \boldsymbol{S}_{\mathcal{T}}\times V_{\mathcal{T}}\times V_{\mathcal{F},0}$ such that the following holds true:

$$(\boldsymbol{\sigma}_T, \boldsymbol{\tau}_T)_{\boldsymbol{L}^2(T)} - (u_T, \nabla{\cdot}\boldsymbol{\tau}_T)_{L^2(T)} + (\lambda_{\partial T}, \boldsymbol{\tau}_T{\cdot}\boldsymbol{n}_T)_{L^2(\partial T)} = 0, \tag{1.52}$$

$$-(\boldsymbol{\sigma}_T, \nabla w_T)_{\boldsymbol{L}^2(T)} + (\boldsymbol{\phi}_{\partial T}{\cdot}\boldsymbol{n}_T, w_T)_{L^2(\partial T)} = (f, w_T)_{L^2(T)}, \tag{1.53}$$

$$([\![\boldsymbol{\phi}_{\partial\mathcal{T}}]\!]{\cdot}\boldsymbol{n}_F, \mu_F)_{L^2(F)} = 0, \tag{1.54}$$

for all $(\boldsymbol{\tau}_T, w_T, \mu_F) \in \boldsymbol{S}_T\times V_T\times V_F$, all $T \in \mathcal{T}$, and all $F \in \mathcal{F}^\circ$, with $\lambda_{\partial T} := (\lambda_F)_{F\in\mathcal{F}_T}$, the HDG numerical flux trace $\boldsymbol{\phi}_{\partial\mathcal{T}} := (\boldsymbol{\phi}_{\partial T})_{T\in\mathcal{T}}$ such that

$$\boldsymbol{\phi}_{\partial T} := \boldsymbol{\sigma}_{T|\partial T} + s^{\mathrm{HDG}}_{\partial T}(u_{T|\partial T} - \lambda_{\partial T})\boldsymbol{n}_T, \quad \forall T \in \mathcal{T}, \tag{1.55}$$

the normal jump across the interface $F = \partial T_- \cap \partial T_+ \cap H_F \in \mathcal{F}^\circ$ defined by

$$[\![\boldsymbol{\phi}_{\partial\mathcal{T}}]\!]{\cdot}\boldsymbol{n}_F := \big(\boldsymbol{\phi}_{\partial T_-|F} - \boldsymbol{\phi}_{\partial T_+|F}\big){\cdot}\boldsymbol{n}_F = \big(\boldsymbol{\phi}_{\partial T_-}{\cdot}\boldsymbol{n}_{T_-}\big)_{|F} + \big(\boldsymbol{\phi}_{\partial T_+}{\cdot}\boldsymbol{n}_{T_+}\big)_{|F}, \tag{1.56}$$

(i.e., $\boldsymbol{n}_F := \boldsymbol{n}_{T_-|F} = -\boldsymbol{n}_{T_+|F}$), and finally, $s_{\partial T}^{\mathrm{HDG}}$ is a linear stabilization operator (to be specified). The Eq. (1.52) is the discrete counterpart of $\boldsymbol{\sigma} = -\nabla u$, the Eq. (1.53) that of $\nabla\cdot\boldsymbol{\sigma} = f$, and the Eq. (1.54) weakly enforces the continuity of the normal component of the numerical flux trace across the mesh interfaces.

Within the above setting, HDG methods are realized by choosing the local spaces $\boldsymbol{S}_T$, V_T, V_F, and the HDG stabilization operator $s_{\partial T}$. Following [62], let us apply this paradigm to the HHO method.

Proposition 1.12 (HHO as HDG method) *The HHO method studied above is rewritten as an HDG method by taking*

$$\boldsymbol{S}_T := \nabla\mathbb{P}_d^{k+1}(T), \qquad V_T := \mathbb{P}_d^k(T), \qquad V_F := \mathbb{P}_{d-1}^k(F), \tag{1.57}$$

and the HDG stabilization operator

$$s_{\partial T}^{\mathrm{HDG}}(v) := h_T^{-1}(\tilde{S}_{\partial T}^* \circ \tilde{S}_{\partial T})(v), \quad \forall v \in V_{\partial T} := \bigtimes_{F\in\mathcal{F}_T} V_F. \tag{1.58}$$

The HDG dual variable is then $\boldsymbol{\sigma}_T = -\nabla R_T(\hat{u}_T)$, the HDG trace variable is $\lambda_{\partial T} = u_{\partial T}$, and the HDG numerical flux trace satisfies $\boldsymbol{\phi}_{\partial T}\cdot\boldsymbol{n}_T = \phi_{\partial T}(\hat{u}_T)$ for all $T \in \mathcal{T}$.

Proof Owing to the choice (1.57) for the local spaces and the definition of the reconstruction operator, (1.52) can be rewritten $(\boldsymbol{\sigma}_T + \nabla R_T(u_T, \lambda_{\partial T}), \boldsymbol{\tau}_T)_{L^2(T)} = 0$ for all $\boldsymbol{\tau}_T \in \boldsymbol{S}_T$. Since $\boldsymbol{\sigma}_T + \nabla R_T(u_T, \lambda_{\partial T}) \in \boldsymbol{S}_T$, this implies that $\boldsymbol{\sigma}_T = -\nabla R_T(u_T, \lambda_{\partial T})$. The rest of the proof is a direct consequence of the identities derived in Proposition 1.11. □

HHO methods were devised independently of HDG methods by adopting the primal viewpoint outlined in Sects. 1.3–1.4, i.e., without introducing a dual variable explicitly. The analysis of HHO methods (see the next chapter) relies on the approximation properties of L^2-orthogonal and elliptic projections, whereas the analysis of HDG methods generally invokes a specific projection operator using Raviart–Thomas finite elements [65] (see also [87]). Furthermore, the HHO stabilization operator from [77, 79] did not have, at the time of its introduction, a counterpart in the setting of HDG methods. Indeed, this operator uses the reconstruction operator, so that at any point $\boldsymbol{x} \in \partial T$, $s_{\partial T}(v)(\boldsymbol{x})$ depends on the values taken by v over the whole boundary ∂T. Instead, the HDG stabilization operator often acts pointwise, that is, $s_{\partial T}(v)(\boldsymbol{x})$ only depends on the value taken by v at $\boldsymbol{x}$. The HHO stabilization operator delivers optimal error estimates for all $k \geq 0$ even on polyhedral meshes. Achieving this result for HDG methods with a stabilization operator acting pointwise requires a subtle design of the local spaces, as explored for instance in [66]. The Lehrenfeld–Schöberl stabilization [115, 116] for HDG+ methods (where the cell unknowns are one degree higher than the face unknowns) is of different nature since $s_{\partial T}(v)(\boldsymbol{x})$ depends on the values taken by v on the face containing $\boldsymbol{x}$. This operator is considered in the context of HHO methods in Sect. 3.2.1.

Remark 1.13 (*Weak Galerkin*) The weak Galerkin (WG) method introduced in [148, 149] can also be embedded into the setting of HDG methods, as shown in [61, Sect. 6.6]. The gradient of the HHO reconstruction operator is called weak gradient in the WG method (not to be confused with the weak gradient in functional analysis). HHO and WG methods were developed independently. In WG methods, the stabilization operator is often based on plain least-squares penalties. A WG method with Lehrenfeld–Schöberl stabilization was considered in [125]. □

1.6 One-Dimensional Setting

This section briefly outlines the HHO method in 1D. The model problem is then $-u'' = f$ in $\Omega := (a, b)$ with the boundary conditions $u(a) = u(b) = 0$. We enumerate the mesh vertices as $(x_i)_{0 \le i \le N+1}$ with $x_0 := a, x_{N+1} := b$. Let $T_i := (x_i, x_{i+1})$ be a generic mesh cell of size h_i for all $0 \le i \le N$. In 1D, the HHO method simplifies since the face unknowns reduce to one real number attached to every mesh vertex. Thus, the choice of the polynomial degree is only relevant to the cell unknowns which are denoted by $u_{\mathcal{T}} := (u_i := u_{T_i})_{0 \le i \le N}$ with $u_i \in \mathbb{P}_1^k(T_i)$ for all $0 \le i \le N$. The face unknowns are denoted by $u_{\mathcal{F}} := \lambda := (\lambda_i)_{0 \le i \le N+1}$ with $\lambda_i \in \mathbb{R}$ for all $0 \le i \le N+1$, and $\lambda_0 = \lambda_{N+1} = 0$ owing to the homogeneous Dirichlet boundary condition. We use the (obvious) notation $\lambda \in \mathbb{R}_{0,0}^{N+2}$ for the face unknowns. It is convenient to define the piecewise affine polynomial $\pi_\lambda^1 : \Omega \to \mathbb{R}$ such that $\pi_\lambda^1(x_i) = \lambda_i$ for all $0 \le i \le N+1$.

Let us first consider the case $k = 0$. Then, on the cell T_i, the discrete unknowns are the real number u_i attached to the cell and the two real numbers $(\lambda_i, \lambda_{i+1})$ attached to the two endpoints of the cell. A direct computation shows that $R'_{T_i}(u_i, (\lambda_i, \lambda_{i+1})) = h_i^{-1}(\lambda_{i+1} - \lambda_i)$ and $S_{\partial T_i}(u_i, (\lambda_i, \lambda_{i+1})) = u_i - \frac{1}{2}(\lambda_i + \lambda_{i+1})$ at both endpoints of T_i. The local discrete equations are for all $(v_i)_{0 \le i \le N}$ and all $(\mu_i)_{0 \le i \le N+1}$,

$$\sum_{0 \le i \le N} \Big(h_i^{-1}(\lambda_{i+1} - \lambda_i)(\mu_{i+1} - \mu_i) \\ + 2h_i^{-1}\big(u_i - \tfrac{1}{2}(\lambda_i + \lambda_{i+1})\big)\big(v_i - \tfrac{1}{2}(\mu_i + \mu_{i+1})\big) \Big) = \sum_{0 \le i \le N} h_i \bar{f}_i v_i, \quad (1.59)$$

where $\bar{f}_i$ denotes the mean-value of f over T_i. Taking first $v_i = \frac{1}{2}(\mu_i + \mu_{i+1})$ for all $0 \le i \le N$ leads to

$$\sum_{0 \le i \le N} h_i^{-1}(\lambda_{i+1} - \lambda_i)(\mu_{i+1} - \mu_i) = \sum_{0 \le i \le N} h_i \bar{f}_i \frac{1}{2}(\mu_i + \mu_{i+1}), \quad (1.60)$$

which is nothing but the transmission problem identified in Proposition 1.10. Using the piecewise affine polynomials π^1_λ and π^1_μ, (1.60) can be rewritten as

$$\int_\Omega (\pi^1_\lambda)'(\pi^1_\mu)'\mathrm{d}x = \int_\Omega \Pi^0_{\mathcal{T}}(f)\pi^1_\mu \mathrm{d}x, \quad \forall \mu \in \mathbb{R}^{N+2}_{0,0}, \tag{1.61}$$

where we recall that $\Pi^0_{\mathcal{T}}$ is the L^2-orthogonal projection onto piecewise constant functions. We recognize in (1.61) the usual finite element discretization of the 1D model problem, up to the projection of the source term. The algebraic realization of (1.61) is $\mathsf{A}\Lambda = \mathsf{F}$, where A is the tridiagonal matrix of size N with entries $(-h^{-1}_{i-1}, h^{-1}_{i-1} + h^{-1}_i, -h^{-1}_i)$, $\Lambda \in \mathbb{R}^N$ is the vector formed by the λ_i's at the interior vertices, and $\mathsf{F} \in \mathbb{R}^N$ has components given by $\mathsf{F}_i := \frac{1}{2}(h_{i-1}\bar{f}_{i-1} + h_i\bar{f}_i)$ for all $1 \le i \le N$. Once the λ_i's have been computed, the cell unknowns are recovered from (1.59) by taking arbitrary cell test functions and zero face test functions. This gives $u_i = \frac{1}{2}h_i^2\bar{f}_i + \frac{1}{2}(\lambda_i + \lambda_{i+1})$ for all $0 \le i \le N$.

A remarkable fact for the HHO method in 1D is that the global transmission problem is the same for all $k \ge 1$. Thus, only the way to post-process locally the face unknowns in order to compute the cell unknowns changes if one modifies the polynomial degree.

Proposition 1.14 (Transmission problem in 1D, $k \ge 1$) *For all $k \ge 1$, the global transmission problem is: Find $\lambda \in \mathbb{R}^{N+2}_{0,0}$ such that*

$$\int_\Omega (\pi^1_\lambda)'(\pi^1_\mu)'\mathrm{d}x = \int_\Omega f\pi^1_\mu \mathrm{d}x, \quad \forall \mu \in \mathbb{R}^{N+2}_{0,0}. \tag{1.62}$$

Proof Since $k \ge 1$, we can consider the pair (π^1_μ, μ) as a test function in the HHO method for all $\mu \in \mathbb{R}^{N+2}_{0,0}$. Since the trace of the cell component equals the face component at every mesh vertex, we infer that $R_{T_i}(\pi^1_{\mu|T_i}, (\mu_i, \mu_{i+1})) = \pi^1_{\mu|T_i} \in \mathbb{P}^1_1(T_i)$, and hence (recall that $k \ge 1$), $S_{\partial T_i}(\pi^1_{\mu|T_i}, (\mu_i, \mu_{i+1})) = 0$ for all $0 \le i \le N$. Using these identities in the HHO method and letting $r_i := R_{T_i}(u_i, (\lambda_i, \lambda_{i+1})) \in \mathbb{P}^{k+1}_1(T_i)$ gives

$$\sum_{0\le i\le N} (r_i', (\pi^1_{\mu|T_i})')_{L^2(T_i)} = \int_\Omega f\pi^1_\mu \mathrm{d}x.$$

Since $(\pi^1_{\mu|T_i})'$ is constant, it only remains to show that $r_i(x_i) = \lambda_i$ and $r_i(x_{i+1}) = \lambda_{i+1}$. This is a remarkable property of the reconstruction in 1D for $k \ge 1$. To prove this fact, we observe that the definition of the reconstruction implies that for all $q \in \mathbb{P}^{k+1}_1(T_i)^\perp$,

$$\big(r_i(x_{i+1}) - \lambda_{i+1}\big)q'(x_{i+1}) - \big(r_i(x_i) - \lambda_i\big)q'(x_i) = \int_{T_i} (r_i - u_i)q''\mathrm{d}x.$$

Since $k \geq 1$, we can take any polynomial $q \in \mathbb{P}_1^2(T_i)^\perp$. Recalling that $\int_{T_i}(r_i - u_i)\mathrm{d}x = 0$ by definition, the claim follows by taking $q \in \mathbb{P}_1^2(T_i)^\perp$ such that $q'(x) = h_i^{-1}(x - x_i)$ and then such that $q'(x) = h_i^{-1}(x_{i+1} - x)$ for all $x \in T_i$. □

Remark 1.15 (*Comparison with FEM*) The HHO method with cell unknowns of degree at most k has as many discrete unknowns as the finite element method based on continuous, piecewise polynomials of degree at most $(k + 2)$. This latter method is more efficient to use since it delivers error estimates with one-order higher convergence rate while it is also amenable to static condensation. □

Chapter 2
Mathematical Aspects

The objective of this chapter is to put the HHO method presented in the previous chapter on a firm mathematical ground. In particular, we prove the key stability and convergence results announced in the previous chapter.

2.1 Mesh Regularity and Basic Analysis Tools

In this section, we give an overview of the basic mathematical notions underlying the analysis of HHO methods: mesh regularity, functional and discrete inverse inequalities, and polynomial approximation properties in Sobolev spaces.

2.1.1 Mesh Regularity

Recall that a mesh $\mathcal{T}$ is composed of polyhedral mesh cells, and $h_{\mathcal{T}}$ denotes the mesh size, i.e., the largest diameter of the cells in $\mathcal{T}$. For simplicity, we assume that Ω is a polyhedron in $\mathbb{R}^d$, $d \geq 2$, so that any mesh covers Ω exactly, i.e., there is no error in the geometric representation of the computational domain. We address the case of a domain with a curved boundary in Sect. 3.2.2. We recall that, by assumption, the mesh faces are planar. This property is used to assert that the normal derivative of a d-variate polynomial at a cell boundary is a piecewise $(d-1)$-variate polynomial.

Since we are interested in a convergence process where the meshes are successively refined, we consider a mesh sequence $\mathfrak{T}$, that is, a countable family of meshes such that 0 is the unique accumulation point of $\{h_{\mathcal{T}}\}_{\mathcal{T}\in\mathfrak{T}}$. The notion of shape-regularity of a mesh sequence is crucial when performing the convergence analysis

M. Cicuttin et al., *Hybrid High-Order Methods*,
SpringerBriefs in Mathematics,
https://doi.org/10.1007/978-3-030-81477-9_2

of any discretization method, since it is instrumental to derive fundamental results on polynomial approximation in the mesh cells, as well as various discrete inverse and functional inequalities. In the simple case where every mesh $\mathcal{T} \in \mathfrak{T}$ is composed of simplices (without hanging nodes), the notion of regularity goes back to Ciarlet [57]: the mesh sequence is said to be shape-regular if there exists a shape-regularity parameter $\rho > 0$ such that for all $\mathcal{T} \in \mathfrak{T}$ and all $T \in \mathcal{T}$ with diameter h_T, $\rho h_T \leq r_T$, where r_T denotes the inradius of the simplex T. In the more general case of meshes composed of polyhedral cells, the mesh sequence is said to be shape-regular if (i) any mesh $\mathcal{T} \in \mathfrak{T}$ admits a matching simplicial submesh $\mathcal{S}_{\mathcal{T}}$ such that any cell (or face) of $\mathcal{S}_{\mathcal{T}}$ is a subset of a cell (or at most one face) of $\mathcal{T}$ and (ii) there exists a shape-regularity parameter $\rho > 0$ such that for all $\mathcal{T} \in \mathfrak{T}$ and all $T \in \mathcal{T}$ and all $S \in \mathcal{S}_{\mathcal{T}}$ such that $S \subset T$, we have $\rho h_S \leq r_S$ and $\rho h_T \leq h_S$. The idea of considering a simplicial submesh to define the regularity of a polyhedral mesh sequence is rather natural. It was considered, e.g., in [28, 76] in the context of discontinuous Galerkin methods. This is also the approach followed in the seminal works on HHO methods [77, 79]. We notice that more general approaches are available, for instance to handle meshes with cells having some very small faces [39, 40].

In what follows, it is implicitly understood that any mesh belongs to a shape-regular mesh sequence, and we do not mention explicitly the mesh sequence. For simplicity, the mesh size is then denoted by h. Moreover, we use the symbol C to denote a generic constant whose value can change at each occurrence as long as it is uniform in the mesh sequence, so that it is, in particular, independent of the mesh size h. The value of C can depend on the domain Ω and the regularity assumptions on the exact solution, the underlying polynomial degree (e.g., the one used in the HHO method), and the shape-regularity parameter ρ of the mesh sequence.

2.1.2 *Functional and Discrete Inverse Inequalities*

Let S be a subset of Ω (typically, S is composed of a collection of mesh cells). For any locally integrable function $v : S \to \mathbb{R}$, $\partial^\alpha v$ denotes the weak partial derivative of v with multi-index $\alpha := (\alpha_1, \dots, \alpha_d) \in \mathbb{N}^d$ of length $|\alpha| := \alpha_1 + \dots + \alpha_d$. Let $m \in \mathbb{N}$ and recall the Sobolev space $H^m(S) := \{v \in L^2(S) \mid \partial^\alpha v \in L^2(S),\ \forall \alpha \in \mathbb{N}^d,\ |\alpha| \leq m\}$ equipped with the following norm and seminorm:

$$\|v\|_{H^m(S)} := \Big(\sum_{|\alpha| \leq m} \ell_S^{2|\alpha|} \|\partial^\alpha v\|^2_{L^2(S)} \Big)^{\frac{1}{2}}, \quad |v|_{H^m(S)} := \Big(\sum_{|\alpha| = m} \|\partial^\alpha v\|^2_{L^2(S)} \Big)^{\frac{1}{2}}, \qquad (2.1)$$

where the length scale $\ell_S := \mathrm{diam}(S)$ is introduced to be dimensionally consistent (notice that the norm and seminorm have different scalings). In some cases, we shall also consider Sobolev spaces of fractional order. Let $s = m + \sigma \in (0, +\infty) \setminus \mathbb{N}$ with $m := \lfloor s \rfloor \in \mathbb{N}$ and $\sigma := s - m \in (0, 1)$. We define $H^s(S) := \{v \in$

$H^m(S) \mid |\partial^\alpha v|_{H^\sigma(S)} < +\infty, \forall \alpha \in \mathbb{N}^d, |\alpha| = m\}$ with the Sobolev–Slobodeckij seminorm

$$|w|_{H^\sigma(S)} := \left(\int_S \int_S \frac{|w(\boldsymbol{x}) - w(\boldsymbol{y})|^2}{\|\boldsymbol{x} - \boldsymbol{y}\|_{\ell^2}^{2\sigma+d}} \,\mathrm{d}\boldsymbol{x}\,\mathrm{d}\boldsymbol{y} \right)^{\frac{1}{2}}. \tag{2.2}$$

We equip $H^s(S)$ with the seminorm $|v|_{H^s(S)} := \left(\sum_{|\alpha|=m} |\partial^\alpha v|^2_{H^\sigma(S)}\right)^{\frac{1}{2}}$ and the norm $\|v\|_{H^s(\Omega)} := \left(\|v\|^2_{H^m(S)} + \ell_S^{2s} |v|^2_{H^s(S)}\right)^{\frac{1}{2}}$.

Let us now state two important functional inequalities valid on every mesh cell $T \in \mathcal{T}$: the Poincaré–Steklov inequality (a.k.a. Poincaré inequality; see [92, Remark 3.32] for a discussion on the terminology) and the multiplicative trace inequality.

Lemma 2.1 (Poincaré–Steklov inequality) *There is C_{PS} such that for all $T \in \mathcal{T}$ and all $v \in H^1(T)$,*

$$\|v - \Pi^0_T(v)\|_{L^2(T)} \leq C_{\mathrm{PS}} h_T \|\nabla v\|_{L^2(T)}, \tag{2.3}$$

where $\Pi^0_T(v)$ is the mean-value of v over T, i.e., the L^2-orthogonal projection of v onto $\mathbb{P}^0_d(T)$. Moreover, there is C such that for all $s \in (0, 1)$, all $T \in \mathcal{T}$ and all $v \in H^s(T)$,

$$\|v - \Pi^0_T(v)\|_{L^2(T)} \leq C h_T^s |v|_{H^s(T)}. \tag{2.4}$$

Lemma 2.2 (Multiplicative trace inequality) *There is C such that for all $T \in \mathcal{T}$ and all $v \in H^1(T)$,*

$$\|v\|_{L^2(\partial T)} \leq C\left(h_T^{-\frac{1}{2}} \|v\|_{L^2(T)} + \|v\|^{\frac{1}{2}}_{L^2(T)} \|\nabla v\|^{\frac{1}{2}}_{L^2(T)}\right). \tag{2.5}$$

Moreover, for all $s \in (\frac{1}{2}, 1)$, there is C such that for all $T \in \mathcal{T}$ and all $v \in H^s(T)$,

$$\|v\|_{L^2(\partial T)} \leq C\left(h_T^{-\frac{1}{2}} \|v\|_{L^2(T)} + h_T^{s-\frac{1}{2}} |v|_{H^s(T)}\right). \tag{2.6}$$

The constant C is uniform with respect to s as long as s is bounded away from $\frac{1}{2}$.

Remark 2.3 (*Literature*) If the mesh cell T is a convex set, the Poincaré–Steklov inequality (2.3) holds true with constant $C_{\mathrm{PS}} = \frac{1}{\pi}$ [13, 132]. In the general case, one decomposes T into the subsimplices resulting from the shape-regularity assumption on the mesh. We refer the reader to [145, Sect. 2.3] and [91, Lemma 5.7] for proofs of Poincaré–Steklov inequalities on composite elements and to [91, Lemma 7.1] for the fractional Poincaré–Steklov inequality (2.4). The idea behind the proof of the multiplicative trace inequality (2.5) in a simplex is to lift the trace using the lowest-order Raviart–Thomas polynomial associated with thc face in question (see [124, Appendix B] and [42, Theorem 4.1]). In a polyhedral cell, for each subface composing ∂T, one carves a subsimplex inside T having equivalent height (see, e.g., [76, Lemma 1.49]). For the fractional multiplicative trace inequality, one considers a pullback to the reference simplex if T is a simplex [91, Lemma 7.2], and one considers a subsimplex as above if T is polyhedral. □

In contrast to functional inequalities, discrete inverse inequalities are only valid in polynomial spaces, and their proof hinges on norm equivalence in a finite-dimensional space. For this reason, discrete inverse inequalities are proven first on a reference simplex and then a geometric mapping is invoked to pass to a generic mesh simplex. In the case of a polyhedral mesh cell, one exploits its decomposition into a finite number of subsimplices; we refer the reader, e.g., to [76, Lemmas 1.44 and 1.46] for more details.

Lemma 2.4 (Discrete inverse inequalities) *Let $l \in \mathbb{N}$ be the polynomial degree. There is C such that for all $T \in \mathcal{T}$ and all $q \in \mathbb{P}^l_d(T)$,*

$$\|\nabla q\|_{L^2(T)} \le C h_T^{-1} \|q\|_{L^2(T)}, \tag{2.7}$$

$$\|q\|_{L^2(\partial T)} \le C h_T^{-\frac{1}{2}} \|q\|_{L^2(T)}. \tag{2.8}$$

2.1.3 Polynomial Approximation

The last question we need to address is how well it is possible to approximate a given function in some Sobolev space by a polynomial. In the context of HHO methods, it is sufficient to consider the approximation by the L^2-orthogonal projection.

Lemma 2.5 (Approximation by L^2-projection) *Let $l \in \mathbb{N}$ be the polynomial degree. Let Π^l_T be the L^2-orthogonal projection onto $\mathbb{P}^l_d(T)$. There is C such that for all $r \in [0, l+1]$, all $m \in \{0, \dots, \lfloor r \rfloor\}$, all $T \in \mathcal{T}$, and all $v \in H^r(T)$,*

$$|v - \Pi^l_T(v)|_{H^m(T)} \le C h_T^{r-m} |v|_{H^r(T)}. \tag{2.9}$$

Moreover, if $r \in (\frac{1}{2}, l+1]$ and $r \in (\frac{3}{2}, l+1]$, $l \ge 1$, respectively, we have

$$\|v - \Pi^l_T(v)\|_{L^2(\partial T)} \le C h_T^{r-\frac{1}{2}} |v|_{H^r(T)}, \quad \|\nabla(v - \Pi^l_T(v))\|_{L^2(\partial T)} \le C h_T^{r-\frac{3}{2}} |v|_{H^r(T)}. \tag{2.10}$$

Proof The estimate (2.9) can be proved by standard arguments in a simplicial cell and by using the arguments from the proof of [91, Lemma 5.6] in a polyhedral cell (the proof combines the Poincaré–Steklov inequalities from Lemma 2.1 with a polynomial built using mean-values of the derivatives of v in T). Let us prove the first bound in (2.10). If $r \in [1, l+1]$, we invoke the multiplicative trace inequality (2.5) which yields

$$\|v - \Pi^l_T(v)\|_{L^2(\partial T)} \le C\big(h_T^{-\frac{1}{2}} \|v - \Pi^l_T(v)\|_{L^2(T)} + h_T^{\frac{1}{2}} |v - \Pi^l_T(v)|_{H^1(T)}\big).$$

The first term is bounded using (2.9) with $m := 0$ and the second term using (2.9) with $m := 1$ (this is possible since $\lfloor r \rfloor \ge 1$). If instead $r \in (\frac{1}{2}, 1)$, the triangle inequality, the discrete trace inequality (2.8), and the bound $\|\Pi_T^l(v) - \Pi_T^0(v)\|_{L^2(T)} \leq 2\|v - \Pi_T^0(v)\|_{L^2(T)}$ imply that

$$\begin{aligned}\|v - \Pi_T^l(v)\|_{L^2(\partial T)} &\leq \|v - \Pi_T^0(v)\|_{L^2(\partial T)} + \|\Pi_T^l(v) - \Pi_T^0(v)\|_{L^2(\partial T)} \\ &\leq \|v - \Pi_T^0(v)\|_{L^2(\partial T)} + Ch_T^{-\frac{1}{2}}\|v - \Pi_T^0(v)\|_{L^2(T)}.\end{aligned}$$

Invoking the fractional multiplicative trace inequality (2.6) to bound the first term on the right-hand side, the bound (2.10) follows from (2.9) with $l = m := 0$ and $|v - \Pi_T^0(v)|_{H^r(T)} = |v|_{H^r(T)}$. Finally, the proof of the second bound in (2.10) is similar, up to the use of the discrete inverse inequality (2.7) together with (2.8). □

2.2 Stability

Recall that for all $T \in \mathcal{T}$, $\hat{V}_T^k$ is equipped with the H^1-like seminorm $|\hat{v}_T|^2_{\hat{V}_T^k} := \|\nabla v_T\|^2_{L^2(T)} + h_T^{-1}\|v_T - v_{\partial T}\|^2_{L^2(\partial T)}$ for all $\hat{v}_T := (v_T, v_{\partial T}) \in \hat{V}_T^k$ (see (1.23)).

Lemma 2.6 (Stability) *There are $0 < \alpha \leq \omega < +\infty$ such that for all $T \in \mathcal{T}$ and all $\hat{v}_T \in \hat{V}_T^k$,*

$$\alpha|\hat{v}_T|^2_{\hat{V}_T^k} \leq a_T(\hat{v}_T, \hat{v}_T) \leq \omega|\hat{v}_T|^2_{\hat{V}_T^k}, \tag{2.11}$$

recalling that $a_T(\hat{v}_T, \hat{v}_T) = \|\nabla R_T(\hat{v}_T)\|^2_{L^2(T)} + h_T^{-1}\|S_{\partial T}(\hat{v}_T)\|^2_{L^2(\partial T)}$.

Proof Let $\hat{v}_T \in \hat{V}_T^k$ and set $r_T := R_T(\hat{v}_T)$.
(i) Lower bound. Let us first bound $\|\nabla v_T\|_{L^2(T)}$. Taking $q := v_T - \Pi_T^0(v_T)$ in the definition (1.17) of r_T and using the Cauchy–Schwarz inequality leads to

$$\begin{aligned}&\|\nabla v_T\|^2_{L^2(T)} = (\nabla r_T, \nabla v_T)_{L^2(T)} + (v_T - v_{\partial T}, \boldsymbol{n}_T{\cdot}\nabla v_T)_{L^2(\partial T)} \\ &\leq \|\nabla r_T\|_{L^2(T)}\|\nabla v_T\|_{L^2(T)} + h_T^{-\frac{1}{2}}\|v_T - v_{\partial T}\|_{L^2(\partial T)} h_T^{\frac{1}{2}}\|\boldsymbol{n}_T{\cdot}\nabla v_T\|_{L^2(\partial T)}.\end{aligned}$$

Invoking the discrete trace inequality (2.8) to bound $\|\boldsymbol{n}_T{\cdot}\nabla v_T\|_{L^2(\partial T)}$ gives

$$\|\nabla v_T\|_{L^2(T)} \leq C\big(\|\nabla r_T\|_{L^2(T)} + h_T^{-\frac{1}{2}}\|v_T - v_{\partial T}\|_{L^2(\partial T)}\big). \tag{2.12}$$

Let us now bound $h_T^{-1}\|v_T - v_{\partial T}\|_{L^2(\partial T)}$. We have

$$\begin{aligned}&\|\Pi_{\partial T}^k(((I - \Pi_T^k)r_T)_{|\partial T})\|_{L^2(\partial T)} \leq \|(I - \Pi_T^k)r_T\|_{L^2(\partial T)} \\ &\leq Ch_T^{-\frac{1}{2}}\|(I - \Pi_T^k)r_T\|_{L^2(T)} \leq Ch_T^{-\frac{1}{2}}\|(I - \Pi_T^0)r_T\|_{L^2(T)} \leq Ch_T^{\frac{1}{2}}\|\nabla r_T\|_{L^2(T)},\end{aligned} \tag{2.13}$$

owing to the L^2-stability of $\Pi^k_{\partial T}$, the discrete trace inequality (2.8), and the Poincaré–Steklov inequality (2.3) (recall that the value of C can change at each occurrence). Using the definition (1.20) of $S_{\partial T}$ and the fact that $v_{T|\partial T} - v_{\partial T}$ is in $\mathbb{P}^k_{d-1}(\mathcal{F}_T)$, we infer that $v_{T|\partial T} - v_{\partial T} = S_{\partial T}(\hat{v}_T) - \Pi^k_{\partial T}(((I - \Pi^k_T)r_T)_{|\partial T})$. The triangle inequality and (2.13) imply that

$$h_T^{-\frac{1}{2}} \|v_T - v_{\partial T}\|_{L^2(\partial T)} \le h_T^{-\frac{1}{2}} \|S_{\partial T}(\hat{v}_T)\|_{L^2(\partial T)} + C\|\nabla r_T\|_{\boldsymbol{L}^2(T)}.$$

Combining this estimate with (2.12) proves the lower bound in (2.11).
(ii) Upper bound. Using the definition (1.17) of r_T with $q := r_T - \Pi^0_T(r_T)$ leads to $\|\nabla r_T\|^2_{\boldsymbol{L}^2(T)} = (\nabla v_T, \nabla r_T)_{\boldsymbol{L}^2(T)} - (v_T - v_{\partial T}, \boldsymbol{n}_T \cdot \nabla r_T)_{L^2(\partial T)}$. Invoking the Cauchy–Schwarz inequality and the discrete trace inequality (2.8) gives

$$\|\nabla r_T\|_{\boldsymbol{L}^2(T)} \le \|\nabla v_T\|_{\boldsymbol{L}^2(T)} + C h_T^{-\frac{1}{2}} \|v_T - v_{\partial T}\|_{L^2(\partial T)}.$$

Moreover, the triangle inequality and the bound (2.13) imply that

$$\begin{aligned} h_T^{-\frac{1}{2}} \|S_{\partial T}(\hat{v}_T)\|_{L^2(\partial T)} &\le h_T^{-\frac{1}{2}} \|v_T - v_{\partial T}\|_{L^2(\partial T)} + h_T^{-\frac{1}{2}} \|\Pi^k_{\partial T}(((I - \Pi^k_T)r_T)_{|\partial T})\|_{L^2(\partial T)} \\ &\le h_T^{-\frac{1}{2}} \|v_T - v_{\partial T}\|_{L^2(\partial T)} + C\|\nabla r_T\|_{\boldsymbol{L}^2(T)}. \end{aligned}$$

Combining the above bounds proves the upper bound in (2.11). □

2.3 Consistency

Let us first prove that the stabilization operator leads to optimal approximation properties when combined with the reduction operator $\hat{I}^k_T$ defined in (1.13). Recall that $\mathcal{E}^{k+1}_T$ denotes the elliptic projection operator onto $\mathbb{P}^{k+1}_d(T)$ (see Lemma 1.4).

Lemma 2.7 (Approximation property of $S_{\partial T} \circ \hat{I}^k_T$) *There is C such that for all $T \in \mathcal{T}$ and all $v \in H^1(T)$,*

$$h_T^{-\frac{1}{2}} \|S_{\partial T}(\hat{I}^k_T(v))\|_{L^2(\partial T)} \le C\|\nabla(v - \mathcal{E}^{k+1}_T(v))\|_{\boldsymbol{L}^2(T)}, \tag{2.14}$$

i.e., we have $h_T^{-\frac{1}{2}} \|S_{\partial T}(\hat{I}^k_T(v))\|_{L^2(\partial T)} \le C \min_{q \in \mathbb{P}^{k+1}_d(T)} \|\nabla(v - q)\|_{\boldsymbol{L}^2(T)}$.

Proof Let $v \in H^1(T)$ and set $\eta := v - \mathcal{E}^{k+1}_T(v)$. Owing to the definition (1.20) of $S_{\partial T}$, the definition (1.13) of $\hat{I}^k_T$, and since $R_T \circ \hat{I}^k_T = \mathcal{E}^{k+1}_T$ (see Lemma 1.4), we have

$$\begin{aligned} S_{\partial T}(\hat{I}^k_T(v)) &= \Pi^k_{\partial T}\left(\Pi^k_T(v)_{|\partial T} - \Pi^k_{\partial T}(v_{|\partial T}) + ((I - \Pi^k_T)\mathcal{E}^{k+1}_T(v))_{|\partial T}\right) \\ &= \Pi^k_T(\eta)_{|\partial T} - \Pi^k_{\partial T}(\eta_{|\partial T}), \end{aligned}$$

since $\Pi^k_{\partial T}(\Pi^k_T(\eta)_{|\partial T}) = \Pi^k_T(\eta)_{|\partial T}$ and $\Pi^k_{\partial T} \circ \Pi^k_{\partial T} = \Pi^k_{\partial T}$. Invoking the triangle inequality, the L^2-stability of $\Pi^k_{\partial T}$, the discrete trace inequality (2.8), and the L^2-stability of Π^k_T leads to (recall that the value of C can change at each occurrence)

$$
\begin{aligned}
&\|S_{\partial T}(\hat{I}^k_T(v))\|_{L^2(\partial T)} \\
&\le \|\Pi^k_T(\eta)\|_{L^2(\partial T)} + \|\Pi^k_{\partial T}(\eta_{|\partial T})\|_{L^2(\partial T)} \le \|\Pi^k_T(\eta)\|_{L^2(\partial T)} + \|\eta\|_{L^2(\partial T)} \\
&\le Ch_T^{-\frac{1}{2}}\|\Pi^k_T(\eta)\|_{L^2(T)} + \|\eta\|_{L^2(\partial T)} \le Ch_T^{-\frac{1}{2}}\|\eta\|_{L^2(T)} + \|\eta\|_{L^2(\partial T)} \le Ch_T^{\frac{1}{2}}\|\nabla\eta\|_{L^2(T)},
\end{aligned}
$$

where the last bound follows from the multiplicative trace inequality (2.5) and the Poincaré–Steklov inequality (2.3) (since $(\eta, 1)_{L^2(T)} = 0$). This proves the bound (2.14), and the bound using the minimum over $q \in \mathbb{P}^{k+1}_d(T)$ readily follows from the definition of the elliptic projection. □

Loosely speaking, the consistency error is measured by inserting the exact solution into the discrete equations and bounding the resulting truncation error. To realize this operation within HHO methods, the idea is to insert $\hat{I}^k_h(u)$ into the discrete equations, where $\hat{I}^k_h$ is the global reduction operator defined in (1.14). Notice that $\hat{I}^k_h(u) \in \hat{V}^k_{h,0}$ since $u \in H^1_0(\Omega)$. With this tool in hand, we define the consistency error $\delta_h \in (\hat{V}^k_{h,0})'$ as the linear form such that for all $\hat{w}_h \in \hat{V}^k_{h,0}$,

$$
\langle \delta_h, \hat{w}_h \rangle := \ell(w_{\mathcal{T}}) - a_h(\hat{I}^k_h(u), \hat{w}_h). \tag{2.15}
$$

Bounding the consistency error then amounts to bounding the dual norm

$$
\|\delta_h\|_* := \sup_{\hat{w}_h \in \hat{V}^k_{h,0}} \frac{|\langle \delta_h, \hat{w}_h \rangle|}{\|\hat{w}_h\|_{\hat{V}^k_{h,0}}}, \tag{2.16}
$$

where the norm $\|\cdot\|_{\hat{V}^k_{h,0}}$ is defined in (1.35). It is implicitly understood here and in what follows that the argument is nonzero when evaluating the dual norm by means of the supremum. To avoid distracting technicalities, we henceforth assume that the exact solution satisfies $u \in H^{1+r}(\Omega)$, $r > \frac{1}{2}$. This assumption actually follows from elliptic regularity theory (see, e.g., [71, p. 158]). It implies that ∇u can be localized as a single-valued function at every mesh face (see [92, Remark 18.4] and also [93, Sect. 41.5] on how to go beyond this assumption for heterogeneous diffusion problems). For all $T \in \mathcal{T}$ and all $v \in H^{1+r}(T)$, $r > \frac{1}{2}$, we define the local seminorm

$$
|v|_{\sharp,T} := \|\nabla v\|_{L^2(T)} + h_T^{\frac{1}{2}}\|\nabla v\|_{L^2(\partial T)}, \tag{2.17}
$$

as well as the global counterpart

$$
|v|_{\sharp,\mathcal{T}} := \Big(\sum_{T\in\mathcal{T}} |v|^2_{\sharp,T}\Big)^{\frac{1}{2}}, \tag{2.18}
$$

for all $v \in H^{1+r}(\mathcal{T}) := \{v \in L^2(\Omega) \mid v_{|T} \in H^{1+r}(T), \quad \forall T \in \mathcal{T}\}$. Let $\mathcal{E}_{\mathcal{T}}^{k+1} : H^1(\Omega) \to \mathbb{P}_d^{k+1}(\mathcal{T})$ be the global elliptic projection operator such that for all $v \in H^1(\Omega)$,

$$\mathcal{E}_{\mathcal{T}}^{k+1}(v)_{|T} := \mathcal{E}_T^{k+1}(v_{|T}), \quad \forall T \in \mathcal{T}. \tag{2.19}$$

Lemma 2.8 (Bound on consistency error) *Assume that the exact solution satisfies $u \in H^{1+r}(\Omega)$, $r > \frac{1}{2}$. There is C such that*

$$\|\delta_h\|_* \leq C|u - \mathcal{E}_{\mathcal{T}}^{k+1}(u)|_{\sharp,\mathcal{T}}. \tag{2.20}$$

Proof Let $\hat{w}_h \in \hat{V}_{h,0}^k$. Integrating by parts in every mesh cell $T \in \mathcal{T}$, recalling that $f = -\Delta u$, and since $\boldsymbol{n}_T \cdot \nabla u$ is meaningful on every face $F \in \mathcal{F}_T$, we obtain

$$\begin{aligned}
\ell(w_{\mathcal{T}}) &= \sum_{T\in\mathcal{T}} (f, w_T)_{L^2(T)} = \sum_{T\in\mathcal{T}} -(\Delta u, w_T)_{L^2(T)} \\
&= \sum_{T\in\mathcal{T}} \big((\nabla u, \nabla w_T)_{\boldsymbol{L}^2(T)} - (\boldsymbol{n}_T \cdot \nabla u, w_T)_{L^2(\partial T)}\big) \\
&= \sum_{T\in\mathcal{T}} \big((\nabla u, \nabla w_T)_{\boldsymbol{L}^2(T)} - (\boldsymbol{n}_T \cdot \nabla u, w_T - w_{\partial T})_{L^2(\partial T)}\big),
\end{aligned}$$

where we used that $\sum_{T\in\mathcal{T}} (\boldsymbol{n}_T \cdot \nabla u, w_{\partial T})_{L^2(\partial T)} = 0$ since ∇u and $w_{\mathcal{F}}$ are single-valued on the mesh interfaces and $w_{\mathcal{F}}$ vanishes on the boundary faces. Moreover, since $\mathcal{E}_T^{k+1} = R_T \circ \hat{I}_T^k$, using the definition of $R_T(\hat{w}_T)$ (with $q := \mathcal{E}_T^{k+1}(u)$) leads to

$$\begin{aligned}
(\nabla R_T(\hat{I}_T^k(u)), \nabla R_T(\hat{w}_T))_{\boldsymbol{L}^2(T)} &= (\nabla \mathcal{E}_T^{k+1}(u), \nabla R_T(\hat{w}_T))_{\boldsymbol{L}^2(T)} \\
&= (\nabla \mathcal{E}_T^{k+1}(u), \nabla w_T)_{\boldsymbol{L}^2(T)} - (\boldsymbol{n}_T \cdot \nabla \mathcal{E}_T^{k+1}(u), w_T - w_{\partial T})_{L^2(\partial T)}.
\end{aligned}$$

Let us set $\eta_T := u_{|T} - \mathcal{E}_T^{k+1}(u)$. Using the definition of a_T and since $(\nabla \eta_T, \nabla w_T)_{\boldsymbol{L}^2(T)} = 0$ owing to (1.18), we have $\langle \delta_h, \hat{w}_h \rangle = -\sum_{T\in\mathcal{T}} (\mathfrak{T}_{1,T} + \mathfrak{T}_{2,T})$ with

$$\begin{aligned}
\mathfrak{T}_{1,T} &:= (\boldsymbol{n}_T \cdot \nabla \eta_T, w_T - w_{\partial T})_{L^2(\partial T)}, \\
\mathfrak{T}_{2,T} &:= h_T^{-1} (S_{\partial T}(\hat{I}_T^k(u)), S_{\partial T}(\hat{w}_T))_{L^2(\partial T)}.
\end{aligned}$$

The Cauchy–Schwarz inequality and the definition of $|\hat{w}_T|_{\hat{V}_T^k}$ imply that

$$|\mathfrak{T}_{1,T}| \leq \|\nabla \eta_T\|_{\boldsymbol{L}^2(\partial T)} \|w_T - w_{\partial T}\|_{L^2(\partial T)} \leq h_T^{\frac{1}{2}} \|\nabla \eta_T\|_{\boldsymbol{L}^2(\partial T)} |\hat{w}_T|_{\hat{V}_T^k}.$$

Moreover, we have

$$|\mathfrak{T}_{2,T}| \leq h_T^{-\frac{1}{2}} \|S_{\partial T}(\hat{I}_T^k(u))\|_{L^2(\partial T)} h_T^{-\frac{1}{2}} \|S_{\partial T}(\hat{w}_T)\|_{L^2(\partial T)}.$$

The first factor is bounded in Lemma 2.7, and the second one in Lemma 2.6. Hence, $|\mathfrak{T}_{2,T}| \leq C\|\nabla\eta_T\|_{L^2(T)}|\hat{w}_T|_{\hat{V}_T^k}$. Collecting these bounds and summing over the mesh cells proves (2.20). □

2.4 H^1-Error Estimate

To allow for a more compact notation, we consider the broken gradient operator $\nabla_{\mathcal{T}}$ and the global reconstruction operator $R_{\mathcal{T}} : \hat{V}_{h,0}^k \to \mathbb{P}_d^{k+1}(\mathcal{T})$ (see (1.30)). For nonnegative real numbers θ, β and a function $\phi \in H^\beta(\mathcal{T}) := \{\phi \in L^2(\Omega) \mid \phi_{|T} \in H^\beta(T),\ \forall T \in \mathcal{T}\}$, we use the shorthand notation

$$|h^\theta \phi|_{H^\beta(\mathcal{T})} := \Big(\sum_{T\in\mathcal{T}} h_T^{2\theta} |\phi|^2_{H^\beta(T)}\Big)^{\frac{1}{2}}. \tag{2.21}$$

Let us introduce the discrete error

$$\hat{e}_h := (e_{\mathcal{T}}, e_{\mathcal{F}}) := \hat{u}_h - \hat{I}_h^k(u) \in \hat{V}_{h,0}^k, \tag{2.22}$$

so that $e_{\mathcal{T}} = u_{\mathcal{T}} - \Pi_{\mathcal{T}}^k(u)$ and $e_{\mathcal{F}} = u_{\mathcal{F}} - \Pi_{\mathcal{F}}^k(u_{|\mathcal{F}})$.

Lemma 2.9 (Discrete H^1-error estimate) *Let $u \in H_0^1(\Omega)$ be the exact solution and let $\hat{u}_h \in \hat{V}_{h,0}^k$ be the HHO solution solving (1.34). Assume that $u \in H^{1+r}(\Omega)$, $r > \frac{1}{2}$. There is C such that*

$$\|\hat{e}_h\|_{\hat{V}_{h,0}^k} + \|\nabla_{\mathcal{T}} R_{\mathcal{T}}(\hat{e}_h)\|_{L^2(\Omega)} + s_h(\hat{e}_h, \hat{e}_h)^{\frac{1}{2}} \leq C|u - \mathcal{E}_{\mathcal{T}}^{k+1}(u)|_{\sharp,\mathcal{T}}. \tag{2.23}$$

Proof We have $a_h(\hat{e}_h, \hat{e}_h) = \langle \delta_h, \hat{e}_h \rangle$, where δ_h is the consistency error defined in (2.15). Summing the lower bound in Lemma 2.6 over all the mesh cells yields

$$\alpha\|\hat{e}_h\|^2_{\hat{V}_{h,0}^k} \leq a_h(\hat{e}_h, \hat{e}_h) = \langle \delta_h, \hat{e}_h \rangle \leq \|\delta_h\|_* \|\hat{e}_h\|_{\hat{V}_{h,0}^k}.$$

Hence, $\|\hat{e}_h\|_{\hat{V}_{h,0}^k} \leq \frac{1}{\alpha}\|\delta_h\|_*$, and Lemma 2.8 yields $\|\hat{e}_h\|_{\hat{V}_{h,0}^k} \leq C|u - \mathcal{E}_{\mathcal{T}}^{k+1}(u)|_{\sharp,\mathcal{T}}$. Finally, (2.23) follows from $a_h(\hat{e}_h, \hat{e}_h) = \|\nabla_{\mathcal{T}} R_{\mathcal{T}}(\hat{e}_h)\|^2_{L^2(\Omega)} + s_h(\hat{e}_h, \hat{e}_h)$ and the above bounds on $a_h(\hat{e}_h, \hat{e}_h)$, $\|\delta_h\|_*$, and $\|\hat{e}_h\|_{\hat{V}_{h,0}^k}$. □

Theorem 2.10 (H^1-error estimate) *Under the assumptions of Lemma 2.9, there is C such that*

$$\|\nabla_{\mathcal{T}}(u - R_{\mathcal{T}}(\hat{u}_h))\|_{L^2(\Omega)} + s_h(\hat{u}_h, \hat{u}_h)^{\frac{1}{2}} \leq C|u - \mathcal{E}_{\mathcal{T}}^{k+1}(u)|_{\sharp,\mathcal{T}}. \tag{2.24}$$

Moreover, if $u \in H^{t+1}(\mathcal{T})$ for some $t \in (\frac{1}{2}, k+1]$, we have

$$\|\nabla_{\mathcal{T}}(u - R_{\mathcal{T}}(\hat{u}_h))\|_{L^2(\Omega)} + s_h(\hat{u}_h, \hat{u}_h)^{\frac{1}{2}} \leq C|h^t u|_{H^{t+1}(\mathcal{T})}. \tag{2.25}$$

This estimate is optimal when $t = k+1$ and converges at rate $O(h^{k+1})$.

Proof (i) The estimate (2.24) follows from (2.23) and the triangle inequality. Indeed, since $R_{\mathcal{T}} \circ \hat{I}_h^k = \mathcal{E}_{\mathcal{T}}^{k+1}$, we have

$$\begin{aligned}
&\|\nabla_{\mathcal{T}}(u - R_{\mathcal{T}}(\hat{u}_h))\|_{L^2(\Omega)} \leq \|\nabla_{\mathcal{T}}(u - \mathcal{E}_{\mathcal{T}}^{k+1}(u))\|_{L^2(\Omega)} + \|\nabla_{\mathcal{T}} R_{\mathcal{T}}(\hat{e}_h)\|_{L^2(\Omega)},\\
&s_h(\hat{u}_h, \hat{u}_h)^{\frac{1}{2}} \leq s_h(\hat{I}_h^k(u), \hat{I}_h^k(u))^{\frac{1}{2}} + s_h(\hat{e}_h, \hat{e}_h)^{\frac{1}{2}},
\end{aligned}$$

$\|\nabla_{\mathcal{T}} R_{\mathcal{T}}(\hat{e}_h)\|_{L^2(\Omega)}$ and $s_h(\hat{e}_h, \hat{e}_h)^{\frac{1}{2}}$ are bounded in Lemma 2.9, and $s_h(\hat{I}_h^k(u), \hat{I}_h^k(u))^{\frac{1}{2}}$ is bounded in Lemma 2.7.
(ii) The estimate (2.25) results from (2.24) and the approximation properties of the local elliptic projection. Indeed, let us set $\eta := u - \mathcal{E}_{\mathcal{T}}^{k+1}(u)$. Owing to the optimality property of the local elliptic projection in the H^1-seminorm and to the approximation property (2.9) of Π_T^{k+1} (with $l := k+1$, $r := 1+t$, $m := 1$, so that $r \leq l+1$ since $t \leq k+1$), we have for all $T \in \mathcal{T}$,

$$\|\nabla\eta\|_{L^2(T)} \leq \|\nabla(u - \Pi_T^{k+1}(u))\|_{L^2(T)} \leq Ch_T^t|u|_{H^{t+1}(T)}.$$

Using the same arguments together with the triangle inequality, the approximation property (2.10), and the discrete trace inequality (2.8), we infer that

$$\begin{aligned}
h_T^{\frac{1}{2}}\|\nabla\eta\|_{L^2(\partial T)} &\leq h_T^{\frac{1}{2}}\|\nabla(u - \Pi_T^{k+1}(u))\|_{L^2(\partial T)} + h_T^{\frac{1}{2}}\|\nabla(\mathcal{E}_T^{k+1}(u) - \Pi_T^{k+1}(u))\|_{L^2(\partial T)}\\
&\leq C\big(h_T^t|u|_{H^{t+1}(T)} + \|\nabla(\mathcal{E}_T(u) - \Pi_T^{k+1}(u))\|_{L^2(T)}\big)\\
&\leq C\big(h_T^t|u|_{H^{t+1}(T)} + 2\|\nabla(u - \Pi_T^{k+1}(u))\|_{L^2(T)}\big) \leq Ch_T^t|u|_{H^{t+1}(T)}.
\end{aligned}$$

We conclude by squaring and summing over the mesh cells. □

2.5 Improved L^2-Error Estimate

As is classical with elliptic problems, an error estimate with a higher-order convergence rate can be established on the L^2-norm of the error. To this purpose, one uses that there are a constant C_{ell} and a regularity pickup index $s \in (\frac{1}{2}, 1]$ such that for all $g \in L^2(\Omega)$, the unique function $\zeta_g \in H_0^1(\Omega)$ such that $a(v, \zeta_g) = (v, g)_{L^2(\Omega)}$ for all $v \in H_0^1(\Omega)$ satisfies the regularity estimate

$$\|\zeta_g\|_{H^{1+s}(\Omega)} \leq C_{\text{ell}}\ell_\Omega^2\|g\|_{L^2(\Omega)}, \tag{2.26}$$

where the scaling factor $\ell_\Omega := \mathrm{diam}(\Omega)$ is introduced to make the constant C_{ell} dimensionless (recall that $\|{\cdot}\|_{H^{1+s}(\Omega)}$ and $\|{\cdot}\|_{L^2(\Omega)}$ have the same scaling and that $-\Delta\zeta_g = g$). The elliptic regularity property (2.26) holds true for the Poisson model problem posed in a polyhedron (see [99, Chap. 4], [71, p. 158]).

Lemma 2.11 (Discrete L^2-error estimate) *Let $u \in H_0^1(\Omega)$ be the exact solution and let $\hat{u}_h \in \hat{V}_{h,0}^k$ be the HHO solution solving (1.34). Assume that $u \in H^{1+r}(\Omega)$, $r > \frac{1}{2}$. Let $s \in (\frac{1}{2}, 1]$ be the pickup index in the elliptic regularity property. Let $\delta := s$ if $k = 0$ and $\delta := 0$ if $k \geq 1$. There is C such that*

$$\|e_{\mathcal{T}}\|_{L^2(\Omega)} \leq C\ell_\Omega^{1-s} h^s \Big(|u - \mathcal{E}_{\mathcal{T}}^{k+1}(u)|_{\sharp,\mathcal{T}} + \ell_\Omega^\delta \|h^{1-\delta}(f - \Pi_{\mathcal{T}}^k(f))\|_{L^2(\Omega)} \Big). \quad (2.27)$$

Proof Let $\zeta_e \in H_0^1(\Omega)$ be such that $a(v, \zeta_e) = (v, e_{\mathcal{T}})_{L^2(\Omega)}$ for all $v \in H_0^1(\Omega)$. Since $-\Delta\zeta_e = e_{\mathcal{T}}$, we have

$$\|e_{\mathcal{T}}\|_{L^2(\Omega)}^2 = -(e_{\mathcal{T}}, \Delta\zeta_e)_{L^2(\Omega)} = \sum_{T\in\mathcal{T}} \big((\nabla e_T, \nabla\zeta_e)_{L^2(T)} + (e_{\partial T} - e_T, \boldsymbol{n}_T{\cdot}\nabla\zeta_e)_{L^2(\partial T)} \big),$$

where we used that $\zeta_e \in H^{1+s}(\Omega)$, $s > \frac{1}{2}$, and $e_F = 0$ for all $F \in \mathcal{F}^\partial$ to infer that $\sum_{T\in\mathcal{T}} (e_{\partial T}, \boldsymbol{n}_T{\cdot}\nabla\zeta_e)_{L^2(\partial T)} = 0$. Let us set $\xi := \zeta_e - \mathcal{E}_{\mathcal{T}}^{k+1}(\zeta_e)$. Adding and subtracting $R_T(\hat{I}_T^k(\zeta_e)) = \mathcal{E}_T^{k+1}(\zeta_e)$ for all $T \in \mathcal{T}$ in the above expression, using the definition of $R_T(\hat{e}_T)$ and since $(\nabla e_T, \nabla\xi)_{L^2(T)} = 0$, we infer that

$$\|e_{\mathcal{T}}\|_{L^2(\Omega)}^2 = \sum_{T\in\mathcal{T}} (e_{\partial T} - e_T, \boldsymbol{n}_T{\cdot}\nabla\xi)_{L^2(\partial T)} + \mathfrak{T}_1,$$

with

$$\mathfrak{T}_1 := \sum_{T\in\mathcal{T}} (\nabla R_T(\hat{e}_T), \nabla R_T(\hat{I}_T^k(\zeta_e)))_{L^2(T)} = -s_h(\hat{e}_h, \hat{I}_h^k(\zeta_e)) + a_h(\hat{e}_h, \hat{I}_h^k(\zeta_e))$$
$$= -s_h(\hat{e}_h, \hat{I}_h^k(\zeta_e)) + (f, \Pi_{\mathcal{T}}^k(\zeta_e))_{L^2(\Omega)} - a_h(\hat{I}_h^k(u), \hat{I}_h^k(\zeta_e)),$$

where we used the definition of $\hat{e}_h$ and the fact that $\hat{u}_h$ solves the HHO problem. Since $(f, \zeta_e)_{L^2(\Omega)} = (\nabla u, \nabla\zeta_e)_{L^2(\Omega)}$, re-arranging the terms leads to $\|e_{\mathcal{T}}\|_{L^2(\Omega)}^2 = \mathfrak{T}_2 + \mathfrak{T}_3 - \mathfrak{T}_4$ with

$$\mathfrak{T}_2 := \sum_{T\in\mathcal{T}} (e_{\partial T} - e_T, \boldsymbol{n}_T{\cdot}\nabla\xi)_{L^2(\partial T)} - s_h(\hat{e}_h, \hat{I}_h^k(\zeta_e)),$$
$$\mathfrak{T}_3 := (\nabla u, \nabla\zeta_e)_{L^2(\Omega)} - a_h(\hat{I}_h^k(u), \hat{I}_h^k(\zeta_e)),$$
$$\mathfrak{T}_4 := (f, \zeta_e - \Pi_{\mathcal{T}}^k(\zeta_e))_{L^2(\Omega)} = (f - \Pi_{\mathcal{T}}^k(f), \zeta_e - \Pi_{\mathcal{T}}^k(\zeta_e))_{L^2(\Omega)}.$$

It remains to bound these three terms. The Cauchy–Schwarz inequality, Lemma 2.6 (to bound $s_h(\hat{e}_h, \hat{e}_h)^{\frac{1}{2}}$), and Lemma 2.7 (to bound $s_h(\hat{I}_h^k(\zeta_e), \hat{I}_h^k(\zeta_e))^{\frac{1}{2}}$) give

$$|\mathfrak{T}_2| \leq C\|\hat{e}_h\|_{\hat{V}_{h,0}^k} |\zeta_e - \mathcal{E}_{\mathcal{T}}^{k+1}(\zeta_e)|_{\sharp,\mathcal{T}}.$$

The approximation property of the elliptic projection gives $|\zeta_e - \mathcal{E}_{\mathcal{T}}^{k+1}(\zeta_e)|_{\sharp,\mathcal{T}} \leq Ch^s|\zeta_e|_{H^{1+s}(\Omega)}$, and $\ell_\Omega^{1+s}|\zeta_e|_{H^{1+s}(\Omega)} \leq \|\zeta_e\|_{H^{1+s}(\Omega)} \leq C_{\text{ell}}\ell_\Omega^2\|e_{\mathcal{T}}\|_{L^2(\Omega)}$ by the elliptic regularity property. Using (2.23) to bound $\|\hat{e}_h\|_{\hat{V}_{h,0}^k}$, we infer that

$$|\mathfrak{T}_2| \leq C|u - \mathcal{E}_{\mathcal{T}}^{k+1}(u)|_{\sharp,\mathcal{T}}\ell_\Omega^{1-s}h^s\|e_{\mathcal{T}}\|_{L^2(\Omega)}.$$

Furthermore, using the definition of a_h, the identity $R_{\mathcal{T}} \circ \hat{I}_h^k = \mathcal{E}_{\mathcal{T}}^{k+1}$, and the orthogonality property of the elliptic projection yields

$$\begin{aligned}\mathfrak{T}_3 &= (\nabla u, \nabla\zeta_e)_{L^2(\Omega)} - (\nabla_{\mathcal{T}}\mathcal{E}_{\mathcal{T}}^{k+1}(u), \nabla_{\mathcal{T}}\mathcal{E}_{\mathcal{T}}^{k+1}(\zeta_e))_{L^2(\Omega)} - s_h(\hat{I}_h^k(u), \hat{I}_h^k(\zeta_e)) \\ &= (\nabla_{\mathcal{T}}(u - \mathcal{E}_{\mathcal{T}}^{k+1}(u)), \nabla_{\mathcal{T}}(\zeta_e - \mathcal{E}_{\mathcal{T}}^{k+1}(\zeta_e)))_{L^2(\Omega)} - s_h(\hat{I}_h^k(u), \hat{I}_h^k(\zeta_e)).\end{aligned}$$

Hence, $|\mathfrak{T}_3| \leq C\|\nabla_{\mathcal{T}}(u - \mathcal{E}_{\mathcal{T}}^{k+1}(u))\|_{L^2(\Omega)}\|\nabla_{\mathcal{T}}(\zeta_e - \mathcal{E}_{\mathcal{T}}^{k+1}(\zeta_e))\|_{L^2(\Omega)}$ by the Cauchy–Schwarz inequality and Lemma 2.7. Invoking again the approximation property of the elliptic projection and the elliptic regularity property yields

$$|\mathfrak{T}_3| \leq C\|\nabla_{\mathcal{T}}(u - \mathcal{E}_{\mathcal{T}}^{k+1}(u))\|_{L^2(\Omega)}\ell_\Omega^{1-s}h^s\|e_{\mathcal{T}}\|_{L^2(\Omega)}.$$

Finally, we have $\|\zeta_e - \Pi_T^k(\zeta_e)\|_{L^2(T)} \leq Ch_T^{1+\gamma}|\zeta_e|_{H^{1+\gamma}(T)}$ with $\gamma := s - \delta$ (i.e., $\gamma = 0$ if $k = 0$ and $\gamma = s$ if $k \geq 1$). The Cauchy–Schwarz inequality implies that $|\mathfrak{T}_4| \leq C\|h^{1+\gamma}(f - \Pi_{\mathcal{T}}^k(f))\|_{L^2(\Omega)}|\zeta_e|_{H^{1+\gamma}(\Omega)}$. Invoking the elliptic regularity property yields $|\mathfrak{T}_4| \leq C\ell_\Omega^{1-\gamma}\|h^{1+\gamma}(f - \Pi_{\mathcal{T}}^k(f))\|_{L^2(\Omega)}\|e_{\mathcal{T}}\|_{L^2(\Omega)}$, so that

$$|\mathfrak{T}_4| \leq C\ell_\Omega^\delta\|h^{1-\delta}(f - \Pi_{\mathcal{T}}^k(f))\|_{L^2(\Omega)}\ell_\Omega^{1-s}h^s\|e_{\mathcal{T}}\|_{L^2(\Omega)}.$$

Putting together the bounds on $\mathfrak{T}_2$, $\mathfrak{T}_3$, and $\mathfrak{T}_4$ completes the proof. □

Theorem 2.12 (L^2-error estimate) *Under the assumptions of Lemma 2.11, there is C such that*

$$\|u - R_{\mathcal{T}}(\hat{u}_h)\|_{L^2(\Omega)} \leq Ch|u - \mathcal{E}_{\mathcal{T}}^{k+1}(u)|_{\sharp,\mathcal{T}} + \|e_{\mathcal{T}}\|_{L^2(\Omega)}. \tag{2.28}$$

Moreover, if $u \in H^{t+1}(\mathcal{T})$ *and* $f \in H^\tau(\mathcal{T})$ *for some* $t \in (\frac{1}{2}, k+1]$, *we have*

$$\|e_{\mathcal{T}}\|_{L^2(\Omega)} + \|u - R_{\mathcal{T}}(\hat{u}_h)\|_{L^2(\Omega)} \leq C\ell_\Omega^{1-s}h^s\big(|h^t u|_{H^{t+1}(\mathcal{T})} + \ell_\Omega^\delta|h^{1-\delta+\tau}f|_{H^\tau(\mathcal{T})}\big), \tag{2.29}$$

with $\tau := \max(t - 1 + \delta, 0)$ *(recall that* $\delta := s$ *if* $k = 0$ *and* $\delta := 0$ *if* $k \geq 1$*). This estimate is optimal when* $t = k + 1$ *and* $s = 1$ *and converges at rate* $O(h^{k+2})$.

Proof (i) The triangle inequality and the Poincaré–Steklov inequality (2.3) give

$$\begin{aligned}\|u - R_{\mathcal{T}}(\hat{u}_h)\|_{L^2(\Omega)} &\le \|u - \mathcal{E}_{\mathcal{T}}^{k+1}(u)\|_{L^2(\Omega)} + \|R_{\mathcal{T}}(\hat{u}_h) - \mathcal{E}_{\mathcal{T}}^{k+1}(u)\|_{L^2(\Omega)} \\ &\le Ch\|\nabla_{\mathcal{T}}(u - \mathcal{E}_{\mathcal{T}}^{k+1}(u))\|_{L^2(\Omega)} + \|R_{\mathcal{T}}(\hat{u}_h) - \mathcal{E}_{\mathcal{T}}^{k+1}(u)\|_{L^2(\Omega)}.\end{aligned}$$

Moreover, we have $\|v\|_{L^2(\Omega)} \le Ch\|\nabla_{\mathcal{T}} v\|_{L^2(\Omega)} + \|\Pi_{\mathcal{T}}^0(v)\|_{L^2(\Omega)}$ for all $v \in H^1(\mathcal{T})$, owing to the triangle inequality and the Poincaré–Steklov inequality (2.3). Applying this bound to $v := R_{\mathcal{T}}(\hat{u}_h) - \mathcal{E}_{\mathcal{T}}^{k+1}(u) = R_{\mathcal{T}}(\hat{e}_h)$ and using that $\Pi_{\mathcal{T}}^0(v) = \Pi_{\mathcal{T}}^0(u_{\mathcal{T}} - u)$, we infer that

$$\|R_{\mathcal{T}}(\hat{u}_h) - \mathcal{E}_{\mathcal{T}}^{k+1}(u)\|_{L^2(\Omega)} \le Ch\|\nabla_{\mathcal{T}} R_{\mathcal{T}}(\hat{e}_h)\|_{L^2(\Omega)} + \|e_{\mathcal{T}}\|_{L^2(\Omega)},$$

since $\|\Pi_{\mathcal{T}}^0(u_{\mathcal{T}} - u)\|_{L^2(\Omega)} \le \|\Pi_{\mathcal{T}}^k(u_{\mathcal{T}} - u)\|_{L^2(\Omega)}$ and $\Pi_{\mathcal{T}}^k(u_{\mathcal{T}} - u) = e_{\mathcal{T}}$. Finally, the estimate (2.28) follows by combining the above inequalities, using (2.23) to bound $\|\nabla_{\mathcal{T}} R_{\mathcal{T}}(\hat{e}_h)\|_{L^2(\Omega)}$, and since $\|\nabla_{\mathcal{T}}(u - \mathcal{E}_{\mathcal{T}}^{k+1}(u))\|_{L^2(\Omega)} \le |u - \mathcal{E}_{\mathcal{T}}^{k+1}(u)|_{\sharp,\mathcal{T}}$.
(ii) (2.29) follows from (2.27)–(2.28) and the approximation properties of the elliptic projection and the L^2-orthogonal projection (notice that in all cases, $\tau \in [0, k+1]$). □

Remark 2.13 (*Regularity assumption*) If $k \ge 1$, the regularity assumption on f in Theorem 2.12 is $f \in H^{t-1}(\mathcal{T})$ which is consistent with the assumption $u \in H^{t+1}(\mathcal{T})$ and the fact that $-\Delta u = f$. If $k = 0$ and, say, $t = 1$, the assumptions become $u \in H^2(\mathcal{T})$ and $f \in H^s(\mathcal{T})$, so that some extra regularity on f is required. □

Chapter 3
Some Variants

The goal of this chapter is to explore some variants of the HHO method devised in Chap. 1 and analyzed in Chap. 2. We first study two variants of the gradient reconstruction operator that will turn useful, for instance, when dealing with nonlinear problems in Chaps. 4 and 7. Then, we explore a mixed-order variant of the HHO method that is useful, for instance, to treat domains with a curved boundary. Finally, we bridge the HHO method to the finite element and virtual element viewpoints.

3.1 Variants on Gradient Reconstruction

In this section, we discuss two variants of the gradient reconstruction operator defined in Sect. 1.3.1. Let $k \geq 0$ be the polynomial degree. Recall that for every mesh cell $T \in \mathcal{T}$, letting $\hat{V}_T^k := \mathbb{P}_d^k(T) \times \mathbb{P}_{d-1}^k(\mathcal{F}_T)$, the local reconstruction operator $R_T : \hat{V}_T^k \to \mathbb{P}_d^{k+1}(T)$ is defined such that for all $\hat{v}_T \in \hat{V}_T^k$,

$$(\nabla R_T(\hat{v}_T), \nabla q)_{L^2(T)} = -(v_T, \Delta q)_{L^2(T)} + (v_{\partial T}, \boldsymbol{n}_T \cdot \nabla q)_{L^2(\partial T)}, \tag{3.1}$$

$$(R_T(\hat{v}_T), 1)_{L^2(T)} = (v_T, 1)_{L^2(T)}, \tag{3.2}$$

where (3.1) holds for all $q \in \mathbb{P}_d^{k+1}(T)^\perp := \{q \in \mathbb{P}_d^{k+1}(T) \mid (q, 1)_{L^2(T)} = 0\}$. The gradient is then reconstructed locally as $\nabla R_T(\hat{v}_T) \in \nabla \mathbb{P}_d^{k+1}(T)$.

A first variant is to reconstruct the gradient in the larger space $\mathbf{P}_d^k(T) := \mathbb{P}_d^k(T; \mathbb{R}^d)$. Notice that $\nabla \mathbb{P}_d^{k+1}(T) \subsetneq \mathbf{P}_d^k(T)$ for all $k \geq 1$, whereas $\nabla \mathbb{P}_d^1(T) = \mathbf{P}_d^0(T)$. Although it may be surprising at first sight to reconstruct a gradient in a space that is not composed of curl-free fields, this choice is relevant in the context of nonlinear problems, as highlighted in [72] for Leray–Lions problems and in [1, 26] for nonlinear elasticity. Indeed, looking at the consistency proof in Lemma 2.8, one sees that one exploits locally the definition of the reconstructed gradient of the test function, $\nabla R_T(\hat{w}_T)$, acting against the reconstructed gradient of some interpolate

M. Cicuttin et al., *Hybrid High-Order Methods*,
SpringerBriefs in Mathematics,
https://doi.org/10.1007/978-3-030-81477-9_3

of the exact solution, $\nabla R_T(\hat{I}_T^k(u))$. However, in the nonlinear case, $\nabla R_T(\hat{w}_T)$ acts against some nonlinear transformation of $\nabla R_T(\hat{I}_T^k(u))$, and there is no reason that this transformation preserves curl-free fields. For further mathematical insight using the notion of limit-conformity, we refer the reader to [72, Sect. 4.1].

The devising of the gradient reconstruction operator $\boldsymbol{G}_T : \hat{V}_T^k \to \mathbb{P}_d^k(T)$ follows the same principle as the one for R_T: it is based on integration by parts. Here, $\boldsymbol{G}_T(\hat{v}_T) \in \mathbb{P}_d^k(T)$ is defined such that for all $\hat{v}_T \in \hat{V}_T^k$,

$$(\boldsymbol{G}_T(\hat{v}_T), \boldsymbol{q})_{\boldsymbol{L}^2(T)} = -(v_T, \nabla\cdot\boldsymbol{q})_{L^2(T)} + (v_{\partial T}, \boldsymbol{n}_T\cdot\boldsymbol{q})_{L^2(\partial T)}, \quad \forall \boldsymbol{q} \in \mathbb{P}_d^k(T). \quad (3.3)$$

To compute $\boldsymbol{G}_T(\hat{v}_T)$, it suffices to invert the mass matrix associated with the scalar-valued polynomial space $\mathbb{P}_d^k(T)$ since only the right-hand side changes when computing each Cartesian component of $\boldsymbol{G}_T(\hat{v}_T)$.

Lemma 3.1 (Gradient reconstruction) (i) $\boldsymbol{\Pi}_{\nabla\mathbb{P}_d^{k+1}}(\boldsymbol{G}_T(\hat{v}_T)) = \nabla R_T(\hat{v}_T)$ *for all* $\hat{v}_T \in \hat{V}_T^k$, *where* $\boldsymbol{\Pi}_{\nabla\mathbb{P}_d^{k+1}}$ *is the* $\boldsymbol{L}^2$*-orthogonal projection onto* $\nabla\mathbb{P}_d^{k+1}(T)$. (ii) $\boldsymbol{G}_T(\hat{I}_T^k(v)) = \boldsymbol{\Pi}_T^k(\nabla v)$ *for all* $v \in H^1(T)$, *where* $\boldsymbol{\Pi}_T^k$ *is the* $\boldsymbol{L}^2$*-orthogonal projection onto* $\mathbb{P}_d^k(T)$.

Proof (i) Let $\hat{v}_T \in \hat{V}_T^k$. For all $q \in \mathbb{P}_d^{k+1}(T)^\perp$, since $\nabla q \in \mathbb{P}_d^k(T)$, (3.3) yields

$$(\boldsymbol{G}_T(\hat{v}_T), \nabla q)_{\boldsymbol{L}^2(T)} = -(v_T, \Delta q)_{L^2(T)} + (v_{\partial T}, \boldsymbol{n}_T\cdot\nabla q)_{L^2(\partial T)} = (\nabla R_T(\hat{v}_T), \nabla q)_{\boldsymbol{L}^2(T)},$$

where the second equality follows from (3.1). Since ∇q is arbitrary in $\nabla\mathbb{P}_d^{k+1}(T)$ and $\nabla R_T(\hat{v}_T) \in \nabla\mathbb{P}_d^{k+1}(T)$, this proves that $\boldsymbol{\Pi}_{\nabla\mathbb{P}_d^{k+1}}(\boldsymbol{G}_T(\hat{v}_T)) = \nabla R_T(\hat{v}_T)$.
(ii) Let $v \in H^1(T)$. Since $\hat{I}_T^k(v) = (\Pi_T^k(v), \Pi_{\partial T}^k(v_{|\partial T}))$, (3.3) yields for all $\boldsymbol{q} \in \mathbb{P}_d^k(T)$,

$$\begin{aligned}(\boldsymbol{G}_T(\hat{I}_T^k(v)), \boldsymbol{q})_{\boldsymbol{L}^2(T)} &= -(\Pi_T^k(v), \nabla\cdot\boldsymbol{q})_{L^2(T)} + (\Pi_{\partial T}^k(v_{|\partial T}), \boldsymbol{n}_T\cdot\boldsymbol{q})_{L^2(\partial T)} \\ &= -(v, \nabla\cdot\boldsymbol{q})_{L^2(T)} + (v, \boldsymbol{n}_T\cdot\boldsymbol{q})_{L^2(\partial T)} = (\nabla v, \boldsymbol{q})_{\boldsymbol{L}^2(T)},\end{aligned}$$

where we used $\nabla\cdot\boldsymbol{q} \in \mathbb{P}_d^{k-1}(T) \subset \mathbb{P}_d^k(T)$, $\boldsymbol{n}_T\cdot\boldsymbol{q}_{|\partial T} \in \mathbb{P}_{d-1}^k(\mathcal{F}_T)$, and integration by parts. Since $\boldsymbol{q}$ is arbitrary in $\mathbb{P}_d^k(T)$, this proves that $\boldsymbol{G}_T(\hat{I}_T^k(v)) = \boldsymbol{\Pi}_T^k(\nabla v)$. □

The property (ii) from Lemma 3.1 is the counterpart of the identity $R_T \circ \hat{I}_T^k = \mathcal{E}_T^{k+1}$ (see Lemma 1.4). By inspecting the proofs of Lemma 2.8 and Theorem 2.10, one readily sees that devising the HHO method with the local bilinear form

$$a_T(\hat{v}_T, \hat{w}_T) := (\boldsymbol{G}_T(\hat{v}_T), \boldsymbol{G}_T(\hat{w}_T))_{\boldsymbol{L}^2(T)} + h_T^{-1}(S_{\partial T}(\hat{v}_T), S_{\partial T}(\hat{w}_T))_{L^2(\partial T)}, \quad (3.4)$$

again leads to optimal H^1- and L^2-error estimates.

Another interesting variant on simplicial meshes is to reconstruct the gradient in the even larger Raviart–Thomas space $\mathbb{RT}_d^k(T) := \mathbb{P}_d^k(T) \oplus \boldsymbol{x}\tilde{\mathbb{P}}_d^k(T)$, where $\tilde{\mathbb{P}}_d^k(T)$ is composed of the restriction to T of the homogeneous d-variate polynomials of degree k. Notice that $\mathbb{P}_d^k(T) \subsetneq \mathbb{RT}_d^k(T) \subsetneq \mathbb{P}_d^{k+1}(T)$. Similarly to (3.3), $\boldsymbol{G}_T^{\mathrm{RT}} : \hat{V}_T^k \to \mathbb{RT}_d^k(T)$ is defined such that for all $\hat{v}_T \in \hat{V}_T^k$, $\boldsymbol{G}_T^{\mathrm{RT}}(\hat{v}_T) \in \mathbb{RT}_d^k(T)$ satisfies

$$(\boldsymbol{G}_T^{\mathrm{RT}}(\hat{v}_T), \boldsymbol{q})_{\boldsymbol{L}^2(T)} = -(v_T, \nabla \cdot \boldsymbol{q})_{L^2(T)} + (v_{\partial T}, \boldsymbol{n}_T \cdot \boldsymbol{q})_{L^2(\partial T)}, \quad \forall \boldsymbol{q} \in \mathbb{RT}_d^k(T). \tag{3.5}$$

In practice, $\boldsymbol{G}_T^{\mathrm{RT}}(\hat{v}_T)$ is computed by inverting the mass matrix associated with the space $\mathbb{RT}_d^k(T)$ (it is not possible here to compute the Cartesian components of $\boldsymbol{G}_T^{\mathrm{RT}}(\hat{v}_T)$ separately). Following the seminal idea from [110] in the context of penalty-free discontinuous Galerkin methods, the motivation for reconstructing a gradient using Raviart–Thomas polynomials is that it allows one to discard the stabilization operator in the HHO method on simplicial meshes [1, 75]. Recall the H^1-like seminorm such that $|\hat{v}_T|^2_{\hat{V}_T^k} := \|\nabla v_T\|^2_{\boldsymbol{L}^2(T)} + h_T^{-1}\|v_T - v_{\partial T}\|^2_{L^2(\partial T)}$ for all $\hat{v}_T \in \hat{V}_T^k$.

Lemma 3.2 (Raviart–Thomas gradient reconstruction) (i) $\boldsymbol{\Pi}_T^k(\boldsymbol{G}_T^{\mathrm{RT}}(\hat{v}_T)) = \boldsymbol{G}_T(\hat{v}_T)$ *for all* $\hat{v}_T \in \hat{V}_T^k$. (ii) $\boldsymbol{G}_T^{\mathrm{RT}}(\hat{I}_T^k(v)) = \boldsymbol{\Pi}_T^k(\nabla v)$ *for all* $v \in H^1(T)$. (iii) *Assuming that the mesh belongs to a shape-regular sequence of simplicial meshes, there is* $C > 0$ *such that* $\|\boldsymbol{G}_T^{\mathrm{RT}}(\hat{v}_T)\|_{\boldsymbol{L}^2(T)} \geq C|\hat{v}_T|_{\hat{V}_T^k}$ *for all* $T \in \mathcal{T}$ *and all* $\hat{v}_T \in \hat{V}_T^k$.

Proof (i) follows from $\mathbb{P}_d^k(T) \subset \mathbb{RT}_d^k(T)$, and (ii) is proved by proceeding as in the proof of Lemma 3.1 and observing that $\nabla \cdot \boldsymbol{q} \in \mathbb{P}_d^k(T)$ and $\boldsymbol{n}_T \cdot \boldsymbol{q}_{|\partial T} \in \mathbb{P}_{d-1}^k(\mathcal{F}_T)$ for all $\boldsymbol{q} \in \mathbb{RT}_d^k(T)$ (even if T is not a simplex). Finally, on a simplex, using classical properties of Raviart–Thomas polynomials (see, e.g., [17, 92]), one can show that for all $\hat{v}_T \in \hat{V}_T^k$, there is $\boldsymbol{q}_v \in \mathbb{RT}_d^k(T)$ such that

$$\Pi_{\partial T}^k(\boldsymbol{n}_T \cdot \boldsymbol{q}_{v|\partial T}) = h_T^{-1}(v_{\partial T} - v_{T|\partial T}), \quad \boldsymbol{\Pi}_T^{k-1}(\boldsymbol{q}_v) = \nabla v_T, \quad \|\boldsymbol{q}_v\|_{L^2(T)} \leq C|\hat{v}_T|_{\hat{V}_T^k}.$$

Using the test function $\boldsymbol{q}_v$ in (3.5) and integrating by parts gives

$$\begin{aligned}(\boldsymbol{G}_T^{\mathrm{RT}}(\hat{v}_T), \boldsymbol{q}_v)_{\boldsymbol{L}^2(T)} &= -(v_T, \nabla \cdot \boldsymbol{q}_v)_{L^2(T)} + (v_{\partial T}, \boldsymbol{n}_T \cdot \boldsymbol{q}_v)_{L^2(\partial T)} \\ &= (\nabla v_T, \boldsymbol{q}_v)_{\boldsymbol{L}^2(T)} - (v_T - v_{\partial T}, \boldsymbol{n}_T \cdot \boldsymbol{q}_v)_{L^2(\partial T)} \\ &= \|\nabla v_T\|^2_{\boldsymbol{L}^2(T)} + h_T^{-1}\|v_T - v_{\partial T}\|^2_{L^2(\partial T)} = |\hat{v}_T|^2_{\hat{V}_T^k},\end{aligned}$$

since $\nabla v_T \in \mathbb{P}_d^{k-1}(T)$ and $v_{T|\partial T} - v_{\partial T} \in \mathbb{P}_{d-1}^k(\mathcal{F}_T)$. The Cauchy–Schwarz inequality and the above bound on $\|\boldsymbol{q}_v\|_{L^2(T)}$ finally imply that

$$\begin{aligned}|\hat{v}_T|^2_{\hat{V}_T^k} = (\boldsymbol{G}_T^{\mathrm{RT}}(\hat{v}_T), \boldsymbol{q}_v)_{L^2(T)} &\leq \|\boldsymbol{G}_T^{\mathrm{RT}}(\hat{v}_T)\|_{\boldsymbol{L}^2(T)}\|\boldsymbol{q}_v\|_{\boldsymbol{L}^2(T)} \\ &\leq C\|\boldsymbol{G}_T^{\mathrm{RT}}(\hat{v}_T)\|_{\boldsymbol{L}^2(T)}|\hat{v}_T|_{\hat{V}_T^k},\end{aligned}$$

which proves the assertion (iii). □

The property (iii) from Lemma 3.2 is the cornerstone ensuring the stability of the HHO method on simplicial meshes using the unstabilized bilinear form

$$a_T(\hat{v}_T, \hat{w}_T) := (\boldsymbol{G}_T^{\mathrm{RT}}(\hat{v}_T), \boldsymbol{G}_T^{\mathrm{RT}}(\hat{w}_T))_{L^2(T)}, \tag{3.6}$$

and the property (ii) is key to deliver optimal error estimates. Notice that the property (ii) fails if the gradient is reconstructed locally in the even larger space $\mathbb{P}_d^{k+1}(T)$ since the normal component on ∂T of polynomials in this space does not necessarily belong to $\mathbb{P}_{d-1}^k(\mathcal{F}_T)$. Notice also that the property (iii) can be achieved on polyhedral meshes by considering Raviart–Thomas polynomials on the simplicial submesh of each mesh cell (see [75]). Another possibility pursued in the context of weak Galerkin methods is to reconstruct the gradient in $\mathbb{P}_d^{k+n-1}(T)$ where n is the number of faces of T [154]; however, the energy-error estimate only decays as $O(h^k)$.

3.2 Mixed-Order Variant and Application to Curved Boundaries

In this section, we briefly discuss the possibility of considering cell and face unknowns that are polynomials of different degrees. As an example of application, we show how a mixed-order variant of the HHO method lends itself to the approximation of problems posed on a domain with a curved boundary.

3.2.1 Mixed-Order Variant with Higher Cell Degree

Let $k \geq 0$ be the polynomial degree for the face unknowns. The degree of the cell unknowns is now set to $k' := k + 1$, leading to a mixed-order HHO method (a mixed-order variant with lower cell degree is briefly addressed below). The mixed-order HHO space is then defined as follows:

$$\hat{V}_h^{k+} := V_{\mathcal{T}}^{k'} \times V_{\mathcal{F}}^k, \qquad V_{\mathcal{T}}^{k'} := \bigtimes_{T\in\mathcal{T}} \mathbb{P}_d^{k'}(T), \qquad V_{\mathcal{F}}^k := \bigtimes_{F\in\mathcal{F}} \mathbb{P}_{d-1}^k(F), \tag{3.7}$$

and the local components of a generic member $\hat{v}_h \in \hat{V}_h^{k+}$ associated with a mesh cell $T \in \mathcal{T}$ and its faces are denoted by $\hat{v}_T := (v_T, v_{\partial T}) \in \hat{V}_T^{k+} := \mathbb{P}_d^{k'}(T) \times \mathbb{P}_{d-1}^k(\mathcal{F}_T)$ with $\mathbb{P}_{d-1}^k(\mathcal{F}_T) := \bigtimes_{F\in\mathcal{F}_T} \mathbb{P}_{d-1}^k(F)$. The HHO reduction operators $\hat{I}_T^{k+} : H^1(T) \to \hat{V}_T^{k+}$ and $\hat{I}_h^{k+} : H^1(\Omega) \to \hat{V}_h^{k+}$ are defined such that

$$\hat{I}_T^{k+}(v) := (\Pi_T^{k'}(v), \Pi_{\partial T}^k(v_{|\partial T})), \qquad \hat{I}_h^{k+}(v) := (\Pi_{\mathcal{T}}^{k'}(v), \Pi_{\mathcal{F}}^k(v_{|\mathcal{F}})). \tag{3.8}$$

The local reconstruction operator $R_T^+ : \hat{V}_T^{k+} \to \mathbb{P}_d^{k+1}(T)$ is defined exactly as in (1.15)–(1.16), and one readily verifies that the identity from Lemma 1.4 can be extended to the mixed-order case, i.e., we have $\mathcal{E}_T^{k+1} = R_T^+ \circ \hat{I}_T^{k+}$ on $H^1(T)$.

The main difference between the equal-order and mixed-order versions of the HHO method lies in the stabilization operator. Indeed, its expression is simpler in the mixed-order case and reads for all $T \in \mathcal{T}$ (compare with (1.20)),

$$S^+_{\partial T}(\hat{v}_T) := \Pi^k_{\partial T}(v_{T|\partial T} - v_{\partial T}) = \Pi^k_{\partial T}(v_{T|\partial T}) - v_{\partial T}. \tag{3.9}$$

The local bilinear form $a^+_T : \hat{V}^{k+}_T \times \hat{V}^{k+}_T \to \mathbb{R}$ is defined as

$$a^+_T(\hat{v}_T, \hat{w}_T) := (\nabla R^+_T(\hat{v}_T), \nabla R^+_T(\hat{w}_T))_{L^2(T)} + h_T^{-1}(S^+_{\partial T}(\hat{v}_T), S^+_{\partial T}(\hat{w}_T))_{L^2(\partial T)}, \tag{3.10}$$

and the global bilinear form $a^+_h : \hat{V}^{k+}_h \times \hat{V}^{k+}_h \to \mathbb{R}$ is still assembled by summing the local contributions cellwise. The discrete problem takes a similar form to (1.34):

$$\begin{cases} \text{Find } \hat{u}_h \in \hat{V}^{k+}_{h,0} := V^{k'}_{\mathcal{T}} \times V^k_{\mathcal{F},0} \text{ such that} \\ a^+_h(\hat{u}_h, \hat{w}_h) = \ell(w_{\mathcal{T}}), \quad \forall \hat{w}_h \in \hat{V}^{k+}_{h,0}. \end{cases} \tag{3.11}$$

The cell unknowns can be eliminated locally by static condensation (see Sect. 1.4.2), and by proceeding as in Sect. 1.5.1, one can recover equilibrated fluxes. Recalling Sect. 1.5.2, we observe that the HDG rewriting of the above mixed-order HHO method has been considered by Lehrenfeld and Schöberl [115, 116] (see also [131]) and is often called HDG+ method.

The analysis of the mixed-order HHO method is quite similar to that of the equal-order version, and we only outline the few changes in the analysis.

Lemma 3.3 (Stability) *Let $|\cdot|_{\hat{V}^{k+}_T}$ denote the extension to $\hat{V}^{k+}_T$ of the H^1-like seminorm defined in (1.23). There are $0 < \alpha \le \omega < +\infty$ such that for all $T \in \mathcal{T}$ and all $\hat{v}_T \in \hat{V}^{k+}_T$,*

$$\alpha|\hat{v}_T|^2_{\hat{V}^{k+}_T} \le a^+_T(\hat{v}_T, \hat{v}_T) \le \omega|\hat{v}_T|^2_{\hat{V}^{k+}_T}. \tag{3.12}$$

Proof Only a few adaptations are needed from the proof of Lemma 2.6. For the lower bound, setting $r_T := R^+_T(\hat{v}_T)$, a slightly sharper version of (2.12) is

$$\|\nabla v_T\|_{L^2(T)} \le C\big(\|\nabla r_T\|_{L^2(T)} + h_T^{-\frac{1}{2}}\|\Pi^k_{\partial T}(v_{T|\partial T}) - v_{\partial T}\|_{L^2(\partial T)}\big) \le C a^+_T(\hat{v}_T, \hat{v}_T)^{\frac{1}{2}}.$$

(Recall that the value of C can change at each occurrence.) Moreover, the triangle inequality, the L^2-optimality of $\Pi^k_{\partial T}$, the discrete trace inequality (2.8), and the Poincaré–Steklov inequality (2.3) imply that

$$\begin{aligned} h_T^{-\frac{1}{2}}\|v_T - v_{\partial T}\|_{L^2(\partial T)} &\le h_T^{-\frac{1}{2}}\|v_T - \Pi_{\partial T}^k(v_{T|\partial T})\|_{L^2(\partial T)} + h_T^{-\frac{1}{2}}\|S_{\partial T}^+(\hat{v}_T)\|_{L^2(\partial T)} \\ &\le h_T^{-\frac{1}{2}}\|v_T' - \Pi_{\partial T}^k(v'_{T|\partial T})\|_{L^2(\partial T)} + h_T^{-\frac{1}{2}}\|S_{\partial T}^+(\hat{v}_T)\|_{L^2(\partial T)} \\ &\le h_T^{-\frac{1}{2}}\|v_T'\|_{L^2(\partial T)} + h_T^{-\frac{1}{2}}\|S_{\partial T}^+(\hat{v}_T)\|_{L^2(\partial T)} \\ &\le C\|\nabla v_T\|_{L^2(T)} + h_T^{-\frac{1}{2}}\|S_{\partial T}^+(\hat{v}_T)\|_{L^2(\partial T)}, \end{aligned}$$

where $v_T' := v_T - \Pi_T^0(v_T)$. Combining these estimates proves the lower bound in (3.12). The proof of the upper bound is similar to the one of Step (ii) in Lemma 2.6, i.e., it combines the bounds $\|\nabla r_T\|_{L^2(T)} \le \|\nabla v_T\|_{L^2(T)} + Ch_T^{-\frac{1}{2}}\|v_T - v_{\partial T}\|_{L^2(\partial T)}$ and $h_T^{-\frac{1}{2}}\|S_{\partial T}^+(\hat{v}_T)\|_{L^2(\partial T)} \le C\|\nabla v_T\|_{L^2(T)} + h_T^{-\frac{1}{2}}\|v_T - v_{\partial T}\|_{L^2(\partial T)}$. □

Lemma 3.4 (Consistency) *There is C such that for all $T \in \mathcal{T}$ and all $v \in H^1(T)$,*

$$h_T^{-\frac{1}{2}}\|S_{\partial T}^+(\hat{I}_T^{k+}(v))\|_{L^2(\partial T)} \le C\|\nabla(v - \Pi_T^{k+1}(v))\|_{L^2(T)}. \tag{3.13}$$

Moreover, defining the consistency error as $\langle\delta_h^+, \hat{w}_h\rangle := \ell(w_{\mathcal{T}}) - a_h(\hat{I}_h^{k+}(u), \hat{w}_h)$ for all $\hat{w}_h \in \hat{V}_{h,0}^{k+}$, with the dual norm $\|\delta_h^+\|_* := \sup_{\hat{w}_h \in \hat{V}_{h,0}^{k+}} \frac{|\langle\delta_h^+, \hat{w}_h\rangle|}{\|\hat{w}_h\|_{\hat{V}_{h,0}^{k+}}}$ and the norm $\|\hat{w}_h\|_{\hat{V}_{h,0}^{k+}} := \big(\sum_{T\in\mathcal{T}} |\hat{w}_T|^2_{\hat{V}_T^{k+}}\big)^{\frac{1}{2}}$, letting the seminorm $|\cdot|_{\sharp,\mathcal{T}}$ be defined as in (2.17)–(2.18), and assuming that the exact solution satisfies $u \in H^{1+r}(\Omega)$, $r > \frac{1}{2}$, we have

$$\|\delta_h^+\|_* \le C\big(|u - \mathcal{E}_{\mathcal{T}}^{k+1}(u)|_{\sharp,\mathcal{T}} + \|\nabla_{\mathcal{T}}(u - \Pi_{\mathcal{T}}^{k+1}(u))\|_{L^2(\Omega)}\big). \tag{3.14}$$

Proof By definition, we have $S_{\partial T}^+(\hat{I}_T^{k+}(v)) = \Pi_{\partial T}^k(\Pi_T^{k+1}(v)_{|\partial T}) - \Pi_{\partial T}^k(v_{|\partial T}) = \Pi_{\partial T}^k((\Pi_T^{k+1}(v) - v)_{|\partial T})$. Using the L^2-stability of $\Pi_{\partial T}^k$, the multiplicative trace inequality (2.5), and the Poincaré–Steklov inequality (2.3), we infer that

$$\begin{aligned} h_T^{-\frac{1}{2}}\|S_{\partial T}^+(\hat{I}_T^{k+}(v))\|_{L^2(\partial T)} &= h_T^{-\frac{1}{2}}\|\Pi_{\partial T}^k((\Pi_T^{k+1}(v) - v)_{|\partial T})\|_{L^2(\partial T)} \\ &\le h_T^{-\frac{1}{2}}\|\Pi_T^{k+1}(v) - v\|_{L^2(\partial T)} \le C\|\nabla(v - \Pi_T^{k+1}(v))\|_{L^2(T)}. \end{aligned}$$

This proves (3.13). Finally, the proof of (3.14) is identical to that of Lemma 2.8 except that we now invoke (3.13) instead of Lemma 2.7. □

Using the above stability and consistency results and reasoning as in the proofs of Lemmas 2.9 and 2.11 and of Theorems 2.10 and 2.12 leads to optimally converging H^1- and L^2-error estimates. The statements and proofs are omitted for brevity (the L^2-error estimate does not require a further regularity assumption on f when $k = 0$ if $k' = k + 1$).

Remark 3.5 (*Mixed-order variant with lower cell degree*) As observed in [62], if the face polynomial degree satisfies $k \ge 1$, the cell polynomial degree can also be set to $k' := k - 1$. One advantage is that there are less cell unknowns to eliminate locally by static condensation. However, the stabilization operator must include a correction depending on the local reconstruction operator as in (1.20) (as in the equal-order case). The stability, consistency, and convergence analysis presented in Chap. 2 can be adapted to the mixed-order HHO method with $k' = k - 1$ as well. The only salient difference is that the improved L^2-norm error estimate requires $k \geq 2$ (this fact was not stated in [62]). Interestingly, as shown in [62], the mixed-order HHO method with $k' = k - 1$ can be bridged to the nonconforming virtual element method introduced in [119] and analyzed in [10]. □

3.2.2 *Domains with a Curved Boundary*

The main idea to treat domains with a curved boundary is to consider the mixed-order HHO method with a higher cell degree and to avoid placing unknowns on the boundary faces. Instead, all the terms involving the boundary are evaluated locally by means of the trace of the corresponding cell unknown. Moreover, the boundary condition is enforced weakly by means of a consistent penalty technique inspired by the seminal work of Nitsche [128]. One novelty is that here the consistency term is directly incorporated into the reconstruction operator, thereby avoiding the need for a penalty parameter that has to be large enough. The main ideas behind the HHO method presented in this section were introduced in [35, 36] with a different reconstruction operator and later simplified in [31], where the presentation dealt with the more general case of an elliptic interface problem.

One way to mesh a domain Ω with a curved boundary consists in embedding it into a larger polyhedral domain Ω' and considering a shape-regular sequence of meshes of Ω'. Notice that these meshes are built without bothering about the location of $\partial\Omega$ inside Ω'. Then, from every mesh $\mathcal{T}'$ of Ω', one generates a mesh $\mathcal{T}$ of Ω by dropping the cells in $\mathcal{T}'$ outside Ω, keeping those inside Ω (called the interior cells), and keeping only the part inside Ω of those cells that are cut by the boundary $\partial\Omega$ (producing the so-called boundary cells). With this process, the cells composing $\mathcal{T}$ cover Ω exactly, the interior cells have planar faces, whereas the boundary cells have one curved face lying on $\partial\Omega$ and planar faces lying inside Ω; see Fig. 3.1.

Some adjustments are still necessary to ensure that the basic analysis tools outlined in Sect. 2.1 are available on the mesh $\mathcal{T}$ that has been constructed this way. The difficulty lies in the fact that the original mesh $\mathcal{T}'$ was deployed without taking into account the position of the boundary $\partial\Omega$ which can therefore cut the cells in an arbitrary way. In particular, some boundary cells of $\mathcal{T}$ can be very small, very flat or have an irregular shape. One possible remedy inspired from the work [109] on discontinuous Galerkin methods is to use a local cell-agglomeration procedure for badly cut cells. This procedure essentially ensures that each mesh cell, possibly after local agglomeration, contains a ball with diameter equivalent to its own diameter. It is shown in [35, Lemma 6.4], [31, Sect. 4.3] that this is indeed possible if the mesh is fine enough, and [35, Lemma 3.4], [31, Lemma 3.4] establish that the discrete inverse inequalities (2.7) and (2.8) then hold true, together with a Poincaré–Steklov inequality on discrete functions. Moreover, it is shown in [35, Lemma 6.1] that if the mesh size is small enough with respect to the curvature of the boundary, every boundary cell $T \in \mathcal{T}$ can be embedded into a ball $T^\dagger$ with equivalent diameter such that the following multiplicative trace inequality holds true: There is C such that for all $T \in \mathcal{T}$ and all $v \in H^1(T^\dagger)$ (setting $T^\dagger := T$ for interior cells),

$$\|v\|_{L^2(\partial T)} \leq C\big(h_T^{-\frac{1}{2}}\|v\|_{L^2(T^\dagger)} + \|v\|_{L^2(T^\dagger)}^{\frac{1}{2}}\|\nabla v\|_{L^2(T^\dagger)}^{\frac{1}{2}}\big). \tag{3.15}$$

Notice that part of the ball $T^\dagger$ may lie outside Ω. Assuming that the exact solution is in $H^{t+1}(\Omega)$ with $t \in [1, k+1]$, polynomial approximation is realized by considering the L^2-orthogonal projection on $T^\dagger$ composed with the stable Calderón–Stein extension operator $E : H^{t+1}(\Omega) \to H^{t+1}(\mathbb{R}^d)$. Thus we set

$$J_T^{k+1}(v) := \big(\Pi_{T^\dagger}^{k+1}(E(v)_{|T^\dagger})\big)_{|T} \in \mathbb{P}_d^{k+1}(T), \tag{3.16}$$

for all $T \in \mathcal{T}$ and all $v \in H^{t+1}(\Omega)$. Notice that $J_T^{k+1}(v) = \Pi_T^{k+1}(v)$ if T is an interior cell. Reasoning as in [35, Lemma 5.6] (i.e., using the approximation properties of $\Pi_{T^\dagger}^{k+1}$ and the multiplicative trace inequality (3.15)), one can show that there is C such that for all $T \in \mathcal{T}$ and all $v \in H^{t+1}(\Omega)$, setting $\eta := v - J_T^{k+1}(v)$,

$$\|\eta\|_{L^2(T)} + h_T^{\frac{1}{2}}\|\eta\|_{L^2(\partial T)} + h_T\|\nabla\eta\|_{L^2(T)} + h_T^{\frac{3}{2}}\|\nabla\eta\|_{L^2(\partial T)} \leq C h_T^{t+1}|E(v)|_{H^{t+1}(T^\dagger)}. \tag{3.17}$$

With these tools in hand, we can devise the mixed-order HHO method for domains with curved boundary. For all $T \in \mathcal{T}$, we consider the partition $\partial T = \partial T^\circ \cup \partial T^\partial$

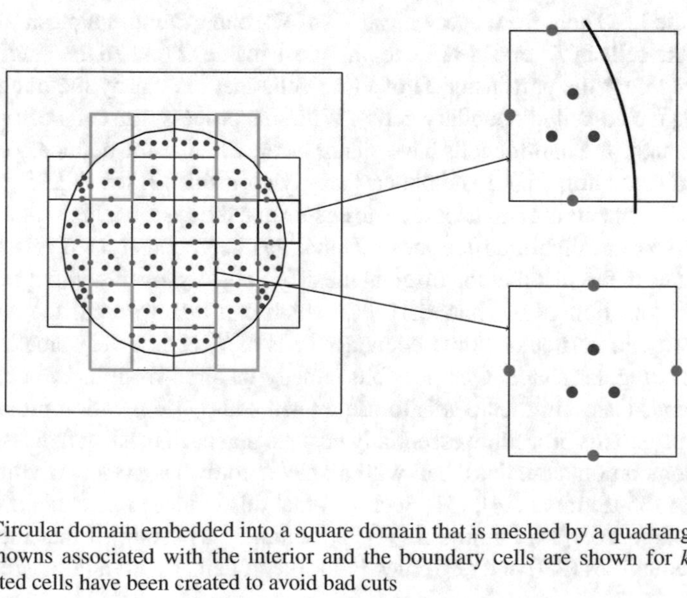

Fig. 3.1 Circular domain embedded into a square domain that is meshed by a quadrangular mesh; HHO unknowns associated with the interior and the boundary cells are shown for $k = 0$; four agglomerated cells have been created to avoid bad cuts

with $\partial T^\circ := \partial T \cap \Omega$, $\partial T^\partial := \partial T \cap \partial\Omega$, and the partition $\mathcal{F}_T = \mathcal{F}_T^\circ \cup \mathcal{F}_T^\partial$ with $\mathcal{F}_T^\circ := \mathcal{F}_T \cap \mathcal{F}^\circ, \mathcal{F}_T^\partial := \mathcal{F}_T \cap \mathcal{F}^\partial$ (the sets $\mathcal{F}^\circ, \mathcal{F}^\partial$ refer to the faces of $\mathcal{T}$ and recall that $T \subset \Omega$). Referring to Fig. 3.1, the mixed-order HHO space is redefined as follows (we keep the same notation for simplicity):

$$\hat{V}_h^{k+} := V_{\mathcal{T}}^{k'} \times V_{\mathcal{F}^\circ}^k, \qquad V_{\mathcal{T}}^{k'} := \bigtimes_{T\in\mathcal{T}} \mathbb{P}_d^{k'}(T), \qquad V_{\mathcal{F}}^k := \bigtimes_{F\in\mathcal{F}^\circ} \mathbb{P}_{d-1}^k(F), \tag{3.18}$$

i.e., $\mathcal{F}^\circ$ is now used in place of $\mathcal{F}$ (no unknowns are attached to the mesh boundary faces in $\mathcal{F}^\partial$). The local components of $\hat{v}_h \in \hat{V}_h^{k+}$ associated with a mesh cell $T \in \mathcal{T}$ and its interior faces are denoted by $\hat{v}_T := (v_T, v_{\partial T}) \in \hat{V}_T^{k+} := \mathbb{P}_d^{k'}(T) \times \mathbb{P}_{d-1}^k(\mathcal{F}_T^\circ)$ with $\mathbb{P}_{d-1}^k(\mathcal{F}_T^\circ) := \bigtimes_{F\in\mathcal{F}_T^\circ} \mathbb{P}_{d-1}^k(F)$. Adapting (1.15), the reconstruction operator $R_T^+ : \hat{V}_T^{k+} \to \mathbb{P}_d^{k+1}(T)$ is such that for all $\hat{v}_T \in \hat{V}_T^{k+}$,

$$(\nabla R_T^+(\hat{v}_T), \nabla q)_{L^2(T)} = -(v_T, \Delta q)_{L^2(T)} + (v_{\partial T}, \boldsymbol{n}_T{\cdot}\nabla q)_{L^2(\partial T^\circ)}, \tag{3.19}$$

$$(R_T^+(\hat{v}_T), 1)_{L^2(T)} = (v_T, 1)_{L^2(T)}, \tag{3.20}$$

where (3.19) holds for all $q \in \mathbb{P}_d^{k+1}(T)^\perp$. Notice that (3.19) is equivalent to

$$\begin{aligned}(\nabla R_T^+(\hat{v}_T), \nabla q)_{L^2(T)} &= (\nabla v_T, \nabla q)_{L^2(T)} - (v_T - v_{\partial T}, \boldsymbol{n}_T{\cdot}\nabla q)_{L^2(\partial T^\circ)} \\ &\quad - (v_T, \boldsymbol{n}_T{\cdot}\nabla q)_{L^2(\partial T^\partial)}.\end{aligned} \tag{3.21}$$

Furthermore, the stabilization operator $S_{\partial T^\circ}^+ : \hat{V}_T^{k+} \to \mathbb{P}_{d-1}^k(\mathcal{F}_T^\circ)$ is such that

$$S_{\partial T^\circ}^+(\hat{v}_T) := \Pi_{\partial T^\circ}^k(v_{T|\partial T^\circ} - v_{\partial T}) = \Pi_{\partial T^\circ}^k(v_{T|\partial T^\circ}) - v_{\partial T}. \tag{3.22}$$

The local bilinear form $a_T^+ : \hat{V}_T^{k+} \times \hat{V}_T^{k+} \to \mathbb{R}$ is defined as

$$\begin{aligned}a_T^+(\hat{v}_T, \hat{w}_T) &:= (\nabla R_T^+(\hat{v}_T), \nabla R_T^+(\hat{w}_T))_{L^2(T)} + h_T^{-1}(S_{\partial T^\circ}^+(\hat{v}_T), S_{\partial T^\circ}^+(\hat{w}_T))_{L^2(\partial T^\circ)} \\ &\quad + h_T^{-1}(v_T, w_T)_{L^2(\partial T^\partial)},\end{aligned} \tag{3.23}$$

where the last term results from the Nitsche's boundary penalty technique. The global bilinear form $a_h^+ : \hat{V}_h^{k+} \times \hat{V}_h^{k+} \to \mathbb{R}$ is still assembled by summing the local contributions cellwise, and the discrete problem seeks $\hat{u}_h \in \hat{V}_h^{k+}$ such that

$$a_h^+(\hat{u}_h, \hat{w}_h) = \ell(w_{\mathcal{T}}), \quad \forall \hat{w}_h \in \hat{V}_h^{k+}. \tag{3.24}$$

The cell unknowns can still be eliminated locally by means of static condensation, and one can again recover equilibrated numerical fluxes. If the Dirichlet condition is non-homogeneous, the right-hand side (3.24) has to be modified to preserve consistency (see [31, 36]).

The error analysis hinges, as usual, on stability and consistency properties. Adapting (1.23), let us equip $\hat{V}_T^{k+}$ with the H^1-like seminorm

$$|\hat{v}_T|^2_{\hat{V}_T^{k+}} := \|\nabla v_T\|^2_{\boldsymbol{L}^2(T)} + h_T^{-1}\|v_T - v_{\partial T}\|^2_{L^2(\partial T^\circ)} + h_T^{-1}\|v_T\|^2_{L^2(\partial T^\partial)}, \tag{3.25}$$

for all $\hat{v}_T \in \hat{V}_T^{k+}$. Proceeding as in the proof of Lemma 3.3, one readily establishes the following stability result.

Lemma 3.6 (Stability) *There are* $0 < \alpha \le \omega < +\infty$ *such that* $\alpha|\hat{v}_T|^2_{\hat{V}_T^{k+}} \le a_T^+(\hat{v}_T, \hat{v}_T) \le \omega|\hat{v}_T|^2_{\hat{V}_T^{k+}}$ *for all* $T \in \mathcal{T}$ *and all* $\hat{v}_T \in \hat{V}_T^{k+}$.

To measure the consistency error, we need to adapt the HHO reduction operator. We now define $\hat{I}_h^{k+}(u) \in \hat{V}_h^{k+}$ such that its local components associated with a mesh cell $T \in \mathcal{T}$ are

$$\hat{I}_T^{k+}(u) := (J_T^{k+1}(u), \Pi^k_{\partial T^\circ}(u_{|\partial T^\circ})) \in \hat{V}_T^{k+}. \tag{3.26}$$

Notice that the cell component of $\hat{I}_T^{k+}(u)$ is now defined using the operator J_T^{k+1} (see (3.16)), whereas the face component (which is now restricted to interior faces) is still defined using an L^2-orthogonal projection on each face. The use of J_T^{k+1} is motivated by the approximation property (3.17), whereas the use of $\Pi^k_{\partial T^\circ}$ is instrumental in the following proofs. Before bounding the consistency error, we need to study the approximation properties of the interpolation operator $\mathcal{J}_T^{k+1} := R_T^+ \circ \hat{I}_T^{k+}$. This operator no longer coincides with the elliptic projection on T because the boundary term in (3.19) is integrated only over ∂T° and because the operator J_T^{k+1} differs from Π_T^{k+1}. Nonetheless, the operator $\mathcal{J}_T^{k+1}$ still enjoys an optimal approximation property.

Lemma 3.7 (Approximation property for $\mathcal{J}_T^{k+1} := R_T^+ \circ \hat{I}_T^{k+}$)*Assume* $u \in H^{t+1}(\Omega)$ *with* $t \in [1, k+1]$. *There is* C *such that for all* $T \in \mathcal{T}$,

$$|u - \mathcal{J}_T^{k+1}(u)|_{\sharp,T} \le C h_T^t |E(u)|_{H^{t+1}(T^\dagger)}, \tag{3.27}$$

recalling that $|v|_{\sharp,T} := \|\nabla v\|_{\boldsymbol{L}^2(T)} + h_T^{\frac{1}{2}}\|\nabla v\|_{\boldsymbol{L}^2(\partial T)}$.

Proof Owing to the triangle inequality and the approximation property (3.17), it suffices to bound $|q|_{\sharp,T}$ with $q := J_T^{k+1}(u) - R_T^+(\hat{I}_T^{k+}(u)) \in \mathbb{P}_d^{k+1}(T)$. Owing to (3.21) and the definition (3.26) of $\hat{I}_T^{k+}$, we infer that

$$\begin{aligned}
\|\nabla q\|^2_{\boldsymbol{L}^2(T)} &= (\nabla(J_T^{k+1}(u) - R_T^+(\hat{I}_T^{k+}(u))), \nabla q)_{\boldsymbol{L}^2(T)} \\
&= -(\Pi^k_{\partial T^\circ}(u) - J_T^{k+1}(u), \boldsymbol{n}_T{\cdot}\nabla q)_{L^2(\partial T^\circ)} + (J_T^{k+1}(u), \boldsymbol{n}_T{\cdot}\nabla q)_{L^2(\partial T^\partial)} \\
&= -(u - J_T^{k+1}(u), \boldsymbol{n}_T{\cdot}\nabla q)_{L^2(\partial T^\circ)} - (u - J_T^{k+1}(u), \boldsymbol{n}_T{\cdot}\nabla q)_{L^2(\partial T^\partial)},
\end{aligned}$$

where we used the L^2-orthogonality property of $\Pi^k_{\partial T^\circ}$ and the fact that the exact solution vanishes on $\partial\Omega$. Invoking the Cauchy–Schwarz inequality, the discrete trace inequality (2.8), and the approximation property (3.17) shows that $\|\nabla q\|_{L^2(T)} \leq Ch_T^t |E(u)|_{H^{t+1}(T^\dagger)}$. Finally, we bound $h_T^{\frac{1}{2}}\|\nabla q\|_{L^2(\partial T)}$ by using the discrete trace inequality (2.8). □

As above, we define the consistency error $\delta_h^+ \in (\hat{V}_h^{k+})'$ as the linear form such that $\langle \delta_h^+, \hat{w}_h\rangle := \ell(w_{\mathcal{T}}) - a_h^+(\hat{I}_h^{k+}(u), \hat{w}_h)$ for all $\hat{w}_h \in \hat{V}_h^{k+}$. We set $\|\delta_h^+\|_* := \sup_{\hat{w}_h \in \hat{V}_h^{k+}} \frac{|\langle \delta_h^+, \hat{w}_h\rangle|}{\|\hat{w}_h\|_{\hat{V}_h^{k+}}}$ with $\|\hat{w}_h\|_{\hat{V}_h^{k+}} := \big(\sum_{T\in\mathcal{T}} |\hat{w}_T|^2_{\hat{V}_T^{k+}}\big)^{\frac{1}{2}}$.

Lemma 3.8 (Consistency) *Assume $u \in H^{t+1}(\Omega)$ with $t \in [1, k+1]$. There is C such that*

$$\|\delta_h^+\|_* \leq C\Big(\sum_{T\in\mathcal{T}} \big(|u - \mathcal{J}_T^{k+1}(u)|^2_{\sharp,T} + h_T^{-1}\|u - J_T^{k+1}(u)\|^2_{L^2(\partial T)}\big)\Big)^{\frac{1}{2}}. \tag{3.28}$$

Proof We have $\langle \delta_h^+, \hat{w}_h\rangle = \mathfrak{T}_1 - \mathfrak{T}_2$ with

$$\mathfrak{T}_1 := \sum_{T\in\mathcal{T}} \big((f, w_T)_{L^2(T)} - (\nabla \mathcal{J}_T^{k+1}(u), \nabla R_T^+(\hat{w}_T))_{L^2(T)}\big),$$

$$\mathfrak{T}_2 := \sum_{T\in\mathcal{T}} \big(h_T^{-1}(S^+_{\partial T^\circ}(\hat{I}_T^{k+}(u)), S^+_{\partial T^\circ}(\hat{w}_T))_{L^2(\partial T^\circ)} + h_T^{-1}(J_T^{k+1}(u), w_T)_{L^2(\partial T^\partial)}\big).$$

Since $f = -\Delta u$, integration by parts and the definition (3.21) of $R_T^+(\hat{w}_T)$ give

$$\begin{aligned}
\mathfrak{T}_1 &= \sum_{T\in\mathcal{T}} \big((\nabla u, \nabla w_T)_{L^2(T)} - (\boldsymbol{n}_T{\cdot}\nabla u, w_T - w_{\partial T})_{L^2(\partial T^\circ)} - (\boldsymbol{n}_T{\cdot}\nabla u, w_T)_{L^2(\partial T^\partial)}\big) \\
&\quad - \sum_{T\in\mathcal{T}} \big((\nabla \mathcal{J}_T^{k+1}(u), \nabla w_T)_{L^2(T)} - (\boldsymbol{n}_T{\cdot}\nabla \mathcal{J}_T^{k+1}(u), w_T - w_{\partial T})_{L^2(\partial T^\circ)} \\
&\quad - (\boldsymbol{n}_T{\cdot}\nabla \mathcal{J}_T^{k+1}(u), w_T)_{L^2(\partial T^\partial)}\big) \\
&= \sum_{T\in\mathcal{T}} \big((\nabla \xi, \nabla w_T)_{L^2(T)} - (\boldsymbol{n}_T{\cdot}\nabla \xi, w_T - w_{\partial T})_{L^2(\partial T^\circ)} - (\boldsymbol{n}_T{\cdot}\nabla \xi, w_T)_{L^2(\partial T^\partial)}\big),
\end{aligned}$$

with $\xi := u - \mathcal{J}_T^{k+1}(u)$, where we used that $\sum_{T\in\mathcal{T}} (\boldsymbol{n}_T{\cdot}\nabla u, w_{\partial T})_{L^2(\partial T^\circ)} = 0$. The term $\mathfrak{T}_1$ is now bounded by using the Cauchy–Schwarz inequality. To bound $\mathfrak{T}_2$, we proceed as in the proof of (3.13) for the interior faces (here, we use again that the face component of $\hat{I}_T^{k+}(u)$ is defined using $\Pi^k_{\partial T^\circ}$), and we use $u_{|\partial T^\partial} = 0$ and the Cauchy–Schwarz inequality for the boundary faces. □

Using the above stability, approximation, and consistency results and reasoning as in the proofs of Lemma 2.9 and Theorem 2.10 leads to optimally converging H^1-error estimates; see [31, Theorem 3.10].

3.3 Finite Element and Virtual Element Viewpoints

Our goal here is to bridge the HHO method with the finite element and virtual element viewpoints. For simplicity, we focus on the equal-order HHO method. Recall that a finite element is defined on a mesh cell $T \in \mathcal{T}$ as a triple (T, P_T, Σ_T), where P_T is a finite-dimensional space composed of functions defined on T and Σ_T are the degrees of freedom, i.e., a collection of linear forms on P_T forming a basis of $\mathcal{L}(P_T; \mathbb{R})$. The material of this section originates from ideas in [60, 62]; see also [117].

Let $k \geq 0$ be the polynomial degree. Consider the finite-dimensional functional space

$$\mathcal{V}_T^k := \{v \in H^1(T) \mid \Delta v \in \mathbb{P}_d^k(T),\ \boldsymbol{n}_T \cdot \nabla v_{|\partial T} \in \mathbb{P}_{d-1}^k(\mathcal{F}_T)\}. \tag{3.29}$$

We observe that $\mathbb{P}_d^{k+1}(T) \subset \mathcal{V}_T^k$, but there are other functions in $\mathcal{V}_T^k$, and these functions are in general not accessible to direct computation. For this reason, the members of $\mathcal{V}_T^k$ are called virtual functions.

Lemma 3.9 ($\mathcal{V}_T^k \leftrightarrow \hat{V}_T^k$) *The linear spaces $\mathcal{V}_T^k$ and $\hat{V}_T^k$ are isomorphic. Consequently,* $\dim(\mathcal{V}_T^k) = \dim(\hat{V}_T^k) = \binom{k+d}{d} + \binom{k+d-1}{d-1}\#\mathcal{F}_T$.

Proof Let us set

$$\begin{aligned}(\hat{V}_T^k)^\perp &:= \{\hat{v}_T \in \hat{V}_T^k \mid (v_T, 1)_{L^2(T)} + (v_{\partial T}, 1)_{L^2(\partial T)} = 0\},\\ (\mathcal{V}_T^k)^\perp &:= \{v \in \mathcal{V}_T^k \mid (v, 1)_{L^2(T)} = 0\}.\end{aligned}$$

To prove the assertion, it suffices to build an isomorphism $\Phi_T : (\hat{V}_T^k)^\perp \to (\mathcal{V}_T^k)^\perp$. For all $\hat{v}_T \in (\hat{V}_T^k)^\perp$, $\varphi := \Phi_T(\hat{v}_T)$ is the unique function in $(\mathcal{V}_T^k)^\perp$ such that $-\Delta\varphi = v_T$ in T and $\boldsymbol{n}_T \cdot \nabla\varphi = v_{\partial T}$ on ∂T. This Neumann problem is well-posed since $(v_T, 1)_{L^2(T)} + (v_{\partial T}, 1)_{L^2(\partial T)} = 0$ and $(\varphi, 1)_{L^2(T)} = 0$. This directly implies that the map Φ_T is bijective. □

We define the virtual reconstruction operator $\mathcal{R}_T : \hat{V}_T^k \to \mathcal{V}_T^k$ such that for all $\hat{v}_T \in \hat{V}_T^k$, the function $\mathcal{R}_T(\hat{v}_T) \in \mathcal{V}_T^k$ is uniquely defined by the following equations:

$$(\nabla\mathcal{R}_T(\hat{v}_T), \nabla w)_{L^2(T)} = -(v_T, \Delta w)_{L^2(T)} + (v_{\partial T}, \boldsymbol{n}_T \cdot \nabla w)_{L^2(\partial T)}, \tag{3.30}$$

$$(\mathcal{R}_T(\hat{v}_T), 1)_{L^2(T)} = (v_T, 1)_{L^2(T)}, \tag{3.31}$$

where (3.30) holds for all $w \in (\mathcal{V}_T^k)^\perp$. Notice that $\mathcal{R}_T(\hat{v}_T)$ is well-defined, but it is not explicitly computable (it can be approximated to a desired accuracy by using, say, a finite element method on a subgrid of T). Let $\hat{\mathcal{J}}_T^k : \mathcal{V}_T^k \to \hat{V}_T^k$ denote the restriction of the reduction operator $\hat{I}_T^k$ to $\mathcal{V}_T^k$. We slightly abuse the terminology by calling the operator $\hat{\mathcal{J}}_T^k$ the degrees of freedom on $\mathcal{V}_T^k$ (one could define more rigorously the degrees of freedom by choosing bases of $\mathbb{P}_d^k(T)$ and $\mathbb{P}_{d-1}^k(F)$ for all $F \in \mathcal{F}_T$). Let I_V denote the identity operator on a generic space V.

Lemma 3.10 (Finite element) *We have $\hat{\mathcal{J}}_T^k \circ \mathcal{R}_T = I_{\hat{V}_T^k}$ and $\mathcal{R}_T \circ \hat{\mathcal{J}}_T^k = I_{\mathcal{V}_T^k}$. Consequently, the triple $(T, \mathcal{V}_T^k, \hat{\mathcal{J}}_T^k)$ is a finite element with interpolation operator $\mathcal{R}_T \circ \hat{I}_T^k$.*

Proof We only need to prove the two identities regarding $\hat{\mathcal{J}}_T^k$ and $\mathcal{R}_T$. (Actually, proving just one identity is sufficient since $\dim(\mathcal{V}_T^k) = \dim(\hat{V}_T^k)$, but we provide two proofs for completeness.)
(i) Let $\hat{v}_T \in \hat{V}_T^k$. To prove that $\hat{\mathcal{J}}_T^k \circ \mathcal{R}_T = I_{\hat{V}_T^k}$, we need to show that

$$\Theta := (\mathcal{R}_T(\hat{v}_T) - v_T, q)_{L^2(T)} + (\mathcal{R}_T(\hat{v}_T) - v_{\partial T}, r)_{L^2(\partial T)} = 0, \quad \forall (q, r) \in \hat{V}_T^k.$$

Let us write $(q, r) = (q', r) + (c, 0)$ with $c := \frac{1}{|T|}\big((q, 1)_{L^2(T)} + (r, 1)_{L^2(\partial T)}\big)$ so that $(q', r) \in (\hat{V}_T^k)^{\perp}$. Using the isomorphism from Lemma 3.9, we set $\psi := \Phi_T(q', r) \in (\mathcal{V}_T^k)^{\perp}$ and observe that

$$\begin{aligned}
\Theta &= (\mathcal{R}_T(\hat{v}_T) - v_T, q')_{L^2(T)} + (\mathcal{R}_T(\hat{v}_T) - v_{\partial T}, r)_{L^2(\partial T)} \\
&= -(\mathcal{R}_T(\hat{v}_T) - v_T, \Delta\psi)_{L^2(T)} + (\mathcal{R}_T(\hat{v}_T) - v_{\partial T}, \boldsymbol{n}_T{\cdot}\nabla\psi)_{L^2(\partial T)} \\
&= (\nabla\mathcal{R}_T(\hat{v}_T), \nabla\psi)_{L^2(T)} + (v_T, \Delta\psi)_{L^2(T)} - (v_{\partial T}, \boldsymbol{n}_T{\cdot}\nabla\psi)_{L^2(\partial T)} = 0,
\end{aligned}$$

where we used (3.31) in the first line, the definition of Φ_T on the second line, and integration by parts and (3.30) on the third line.
(ii) Let $v \in \mathcal{V}_T^k$. The definition (3.30) of $\mathcal{R}_T$ implies that for all $w \in (\mathcal{V}_T^k)^{\perp}$, we have

$$\begin{aligned}
(\nabla\mathcal{R}_T(\hat{\mathcal{J}}_T^k(v)), \nabla w)_{L^2(T)} &= -(\Pi_T^k(v), \Delta w)_{L^2(T)} + (\Pi_{\partial T}^k(v), \boldsymbol{n}_T{\cdot}\nabla w)_{L^2(\partial T)} \\
&= -(v, \Delta w)_{L^2(T)} + (v, \boldsymbol{n}_T{\cdot}\nabla w)_{L^2(\partial T)} = (\nabla v, \nabla w)_{L^2(T)},
\end{aligned}$$

since $\Delta w \in \mathbb{P}_d^k(T)$ and $\boldsymbol{n}_T{\cdot}\nabla w_{|\partial T} \in \mathbb{P}_{d-1}^k(\mathcal{F}_T)$. Since $\mathcal{R}_T(\hat{\mathcal{J}}_T^k(v)) - v \in (\mathcal{V}_T^k)^{\perp}$ owing to (3.31), and w is arbitrary in $(\mathcal{V}_T^k)^{\perp}$, $\mathcal{R}_T(\hat{\mathcal{J}}_T^k(v)) = v$. □

Remark 3.11 (*Right inverse*) Lemma 3.10 implies that $\mathcal{R}_T : \hat{V}_T^k \to \mathcal{V}_T^k$ is a right inverse of $\hat{I}_T^k$. Right inverses with other codomains can be devised. For instance, the right inverse devised in [95] using bubble functions maps onto $\mathbb{P}_d^{k+d+1}(T)$ on simplicial meshes and allows one to build a globally H^1-conforming function. The construction can be extended to general meshes. □

Let us define the high-order Crouzeix–Raviart-type finite element space

$$\mathcal{V}_h^k := \{v_h \in L^2(\Omega) \mid v_{h|T} \in \mathcal{V}_T^k,\ \forall T \in \mathcal{T},\ \Pi_F^k([\![v_h]\!]) = 0,\ \forall F \in \mathcal{F}^{\circ}\}, \quad (3.32)$$

where $[\![v_h]\!]$ denotes the jump of v_h across the mesh interface $F \in \mathcal{F}^{\circ}$, and let us set $\mathcal{V}_{h,0}^k := \{v_h \in \mathcal{V}_h^k \mid \Pi_F^k(v_h) = 0,\ \forall F \in \mathcal{F}^{\partial}\}$. The global degrees of freedom of a function $v_h \in \mathcal{V}_{h,0}^k$ are $\hat{\mathcal{J}}_h^k(v_h) := (\Pi_{\mathcal{T}}^k(v_h), \Pi_{\mathcal{F}}^k(v_{h|\mathcal{F}})) \in \hat{V}_{h,0}^k$, which is meaningful owing to jump condition in (3.32). A natural way to use the finite element identified in Lemma 3.10 to approximate the model problem (1.2) is to seek $u_h \in \mathcal{V}^k_{h,0}$ such that

$$a^{\mathrm{V}}_h(u_h, w_h) = \ell(\Pi^k_{\mathcal{T}}(w_h)), \quad \forall w_h \in \mathcal{V}^k_{h,0}, \tag{3.33}$$

with $a^{\mathrm{V}}_h : \mathcal{V}^k_{h,0} \times \mathcal{V}^k_{h,0} \to \mathbb{R}$ such that $a^{\mathrm{V}}_h(v_h, w_h) := (\nabla_{\mathcal{T}} v_h, \nabla_{\mathcal{T}} w_h)_{\boldsymbol{L}^2(\Omega)}$ (notice that $v_h \mapsto \|\nabla_{\mathcal{T}} v_h\|_{\boldsymbol{L}^2(\Omega)}$ defines a norm on $\mathcal{V}^k_{h,0}$). The use of the L^2-orthogonal projection to evaluate the right-hand side of (3.33) is to stick to the multiscale HHO method proposed in [60]. Using the above reduction and reconstruction operators, an equivalent reformulation of (3.33) is to seek $\hat{u}_h \in \hat{V}^k_{h,0}$ such that $\hat{a}^{\mathrm{V}}_h(\hat{u}_h, \hat{w}_h) = \ell(w_{\mathcal{T}})$ for all $\hat{w}_h \in \hat{V}^k_{h,0}$, with $\hat{a}^{\mathrm{V}}_h(\hat{v}_h, \hat{w}_h) := (\nabla_{\mathcal{T}} \mathcal{R}_{\mathcal{T}}(\hat{v}_h), \nabla_{\mathcal{T}} \mathcal{R}_{\mathcal{T}}(\hat{w}_h))_{\boldsymbol{L}^2(\Omega)}$ and $\mathcal{R}_{\mathcal{T}}(\hat{v}_h)_{|T} := \mathcal{R}_T(\hat{v}_T)$ for all $T \in \mathcal{T}$.

The discrete problem (3.33) is not easily tractable since the computation of the basis functions in $\mathcal{V}^k_{h,0}$ is possible only by using some subgrid discretization method in each mesh cell. This approach is reasonable when dealing with a diffusion problem characterized by subgrid scales that are not captured by the mesh $\mathcal{T}$. Instead, in the absence of multiscale features, it is more efficient to use the original HHO method presented in Chap. 1. To bridge the two methods, we first notice that an equivalent formulation of the original HHO method is to seek $u_h \in \mathcal{V}^k_{h,0}$ such that $a^{\mathrm{HHO}}_h(u_h, w_h) = \ell(\Pi^k_{\mathcal{T}}(w_h))$ for all $w_h \in \mathcal{V}^k_{h,0}$, with $a^{\mathrm{HHO}}_h : \mathcal{V}^k_{h,0} \times \mathcal{V}^k_{h,0} \to \mathbb{R}$ such that

$$a^{\mathrm{HHO}}_h(v_h, w_h) := (\nabla_{\mathcal{T}} \mathcal{E}^{k+1}_{\mathcal{T}}(v_h), \nabla_{\mathcal{T}} \mathcal{E}^{k+1}_{\mathcal{T}}(w_h))_{\boldsymbol{L}^2(\Omega)} + s_h(\hat{\mathcal{J}}^k_h(v_h), \hat{\mathcal{J}}^k_h(w_h)). \tag{3.34}$$

The quantity $a^{\mathrm{HHO}}_h(v_h, w_h)$ is computable even if one only knows the global degrees of freedom $\hat{\mathcal{J}}^k_h(v_h)$ and $\hat{\mathcal{J}}^k_h(w_h)$ of v_h and w_h, without the need to explicitly knowing these functions (notice that $\mathcal{E}^{k+1}_T = R_T \circ \hat{\mathcal{J}}^k_T$ for all $T \in \mathcal{T}$, where R_T is the computable reconstruction operator defined in (1.15)–(1.16)). The role of the stabilization in the original HHO method can then be understood as a computable way of ensuring that a^{HHO}_h remains H^1-coercive on $\mathcal{V}^k_{h,0}$, in the spirit of the seminal ideas developed for the virtual element method (see, e.g., [14]).

Lemma 3.12 (Coercivity on $\mathcal{V}^k_{h,0}$) *There is $\eta > 0$ such that $a^{\mathrm{HHO}}_h(v_h, v_h) \geq \eta \|\nabla_{\mathcal{T}} v_h\|^2_{\boldsymbol{L}^2(\Omega)}$ for all $v_h \in \mathcal{V}^k_{h,0}$.*

Proof (i) Let us first prove the following inverse inequalities: There is C such that for all $T \in \mathcal{T}$ and all $v \in \mathcal{V}^k_T$,

$$\|\boldsymbol{n}_T{\cdot}\nabla v\|_{L^2(\partial T)} + h_T^{\frac{1}{2}} \|\Delta v\|_{L^2(T)} \leq C h_T^{-\frac{1}{2}} \|\nabla v\|_{\boldsymbol{L}^2(T)}. \tag{3.35}$$

Let $F \in \mathcal{F}_T$ and set $q := (\boldsymbol{n}_T{\cdot}\nabla v)_{|F} \in \mathbb{P}^k_{d-1}(F)$. For simplicity, we assume that T is a simplex (otherwise, for every simplicial subface of F, one carves a simplex inside T of diameter uniformly equivalent to h_T). It results from [94, Lemma A.3] that there is C_{DIV} such that for all $q \in \mathbb{P}^k_{d-1}(F)$, there is $\boldsymbol{\theta}_q \in \mathbf{RT}^k_d(T)$ (the Raviart–Thomas finite element space of order k in T) satisfying $\boldsymbol{n}_T{\cdot}\boldsymbol{\theta}_q = q$ on F, $\nabla{\cdot}\boldsymbol{\theta}_q = \Delta v \in \mathbb{P}^k_d(T)$, and

$$\|\boldsymbol{\theta}_q\|_{\boldsymbol{L}^2(T)} \le C_{\mathrm{DIV}} \min_{\boldsymbol{a}\in \boldsymbol{V}_q} \|\boldsymbol{a}\|_{\boldsymbol{L}^2(T)},$$

with $\boldsymbol{V}_q := \{\boldsymbol{a} \in \boldsymbol{H}(\operatorname{div}; T) \mid (\boldsymbol{n}_T{\cdot}\boldsymbol{a})_{|F} = q,\ \nabla{\cdot}\boldsymbol{a} = \Delta v\}$. Since $\nabla v \in \boldsymbol{V}_q$, invoking the discrete trace inequality (2.8) for the polynomial $\boldsymbol{\theta}_q$, we infer that

$$\|\boldsymbol{n}_T{\cdot}\nabla v\|_{L^2(F)} = \|\boldsymbol{n}_T{\cdot}\boldsymbol{\theta}_q\|_{L^2(F)} \le C h_T^{-\frac{1}{2}} \|\boldsymbol{\theta}_q\|_{\boldsymbol{L}^2(T)} \le C C_{\mathrm{DIV}} h_T^{-\frac{1}{2}} \|\nabla v\|_{\boldsymbol{L}^2(T)}. \quad (3.36)$$

Moreover, setting $r := \Delta v \in \mathbb{P}_d^k(T)$, an integration by parts gives $\|\Delta v\|_{L^2(T)}^2 = -(\nabla v, \nabla r)_{\boldsymbol{L}^2(T)} + (\boldsymbol{n}_T{\cdot}\nabla v, r)_{L^2(\partial T)}$. Invoking the Cauchy–Schwarz inequality, the discrete inverse inequalities (2.7)–(2.8), and the bound (3.36) readily gives $\|\Delta v\|_{L^2(T)} \le C h_T^{-1} \|\nabla v\|_{\boldsymbol{L}^2(T)}$. This completes the proof of (3.35).
(ii) Let $v_h \in \mathcal{V}_{h,0}^k$. Integrating by parts in (3.30), invoking the Cauchy–Schwarz inequality, and using (3.35) shows that $\|\nabla \mathcal{R}_T(\hat{v}_T)\|_{\boldsymbol{L}^2(T)} \le C|\hat{v}_T|_{\hat{V}_T^k}$ for all $\hat{v}_T \in \hat{V}_T^k$ and all $T \in \mathcal{T}$ (we only use the bound on the normal derivative in (3.35)). Owing to Lemma 3.10, we infer that $\|\nabla v_{h|T}\|_{\boldsymbol{L}^2(T)} \le C|\hat{\mathcal{J}}_T^k(v_{h|T})|_{\hat{V}_T^k}$. The assertion now follows by invoking the lower bound in Lemma 2.6 and summing cellwise since $a_h^{\mathrm{HHO}}(v_h, v_h) = \sum_{T\in\mathcal{T}} a_T(\hat{\mathcal{J}}_T^k(v_{h|T}), \hat{\mathcal{J}}_T^k(v_{h|T}))$. □

Chapter 4
Linear Elasticity and Hyperelasticity

In this chapter, we show how to discretize using HHO methods linear elasticity and nonlinear hyperelasticity problems. In particular, we pay particular attention to the robustness of the discretization in the quasi-incompressible limit. For linear elasticity, we reconstruct the strain tensor in the space composed of symmetric gradients of vector-valued polynomials. For nonlinear hyperelasticity, we reconstruct the deformation gradient in a full tensor-valued polynomial space, and not just in a space composed of polynomial gradients. We also consider a second gradient reconstruction in an even larger space built using Raviart–Thomas polynomials, for which no additional stabilization is necessary. Finally, we present some numerical examples.

4.1 Continuum Mechanics

We are interested in finding the static equilibrium configuration of an elastic continuum body that occupies the domain Ω in the reference configuration. Here, $\Omega \subset \mathbb{R}^d$, $d \in \{2, 3\}$, is a bounded Lipschitz domain with unit outward normal $\boldsymbol{n}$ and boundary partitioned as $\partial\Omega = \overline{\partial\Omega_N} \cup \overline{\partial\Omega_D}$ with two relatively open and disjoint subsets $\partial\Omega_N$ and $\partial\Omega_D$. The body undergoes deformations under the action of a body force $\boldsymbol{f} : \Omega \to \mathbb{R}^d$, a traction force $\boldsymbol{g}_N : \partial\Omega_N \to \mathbb{R}^d$, and a prescribed displacement $\boldsymbol{u}_D : \partial\Omega_D \to \mathbb{R}^d$. We assume that $\partial\Omega_D$ has positive measure so as to prevent rigid-body motions. Due to the deformation, a point $\boldsymbol{x} \in \Omega$ in the reference configuration is mapped to a point $\boldsymbol{x}' = \boldsymbol{x} + \boldsymbol{u}(\boldsymbol{x})$ in the equilibrium configuration, where $\boldsymbol{u} : \Omega \to \mathbb{R}^d$ is the displacement field.

M. Cicuttin et al., *Hybrid High-Order Methods*,
SpringerBriefs in Mathematics,
https://doi.org/10.1007/978-3-030-81477-9_4

4.1.1 Infinitesimal Deformations and Linear Elasticity

Since we are concerned here with infinitesimal deformations, a relevant measure of the deformations of the body is the linearized strain tensor such that

$$\boldsymbol{\varepsilon}(\boldsymbol{u}) := \frac{1}{2}(\nabla\boldsymbol{u} + \nabla\boldsymbol{u}^{\mathsf{T}}). \tag{4.1}$$

Notice that $\boldsymbol{\varepsilon}(\boldsymbol{u})$ takes values in the space $\mathbb{R}^{d\times d}_{\mathrm{sym}}$ composed of symmetric tensors of order d. Moreover, in the framework of linear isotropic elasticity, the internal stresses in the body are described at any point in Ω by the stress tensor $\boldsymbol{\sigma}$ which depends on the linearized strain tensor $\boldsymbol{\varepsilon}$ at that point. The constitutive stress-strain relation is linear and takes the form

$$\boldsymbol{\sigma}(\boldsymbol{\varepsilon}) := 2\mu\boldsymbol{\varepsilon} + \lambda\,\mathrm{tr}(\boldsymbol{\varepsilon})\boldsymbol{I}_d, \tag{4.2}$$

where λ and μ are material parameters called Lamé coefficients, and $\boldsymbol{I}_d$ is the identity tensor in $\mathbb{R}^{d\times d}$. Notice that $\boldsymbol{\sigma}(\boldsymbol{\varepsilon})$ also takes values in $\mathbb{R}^{d\times d}_{\mathrm{sym}}$ (the symmetry of $\boldsymbol{\sigma}(\boldsymbol{\varepsilon})$ is actually a consequence of the balance of angular momentum in infinitesimal deformations). For simplicity, we assume that λ and μ are constant in Ω. Owing to thermodynamic stability, we have $\mu > 0$ and $\lambda + \frac{2}{3}\mu > 0$. The coefficient $\kappa := \lambda + \frac{2}{3}\mu$, called bulk modulus, describes the compressibility of the material. Very large values relative to μ, i.e., $\lambda \gg \mu$, correspond to almost incompressible materials.

In the above setting, the displacement field $\boldsymbol{u} : \Omega \to \mathbb{R}^d$ satisfies the following equations:

$$-\nabla\cdot\boldsymbol{\sigma}(\boldsymbol{\varepsilon}(\boldsymbol{u})) = \boldsymbol{f} \quad \text{in } \Omega, \tag{4.3}$$

$$\boldsymbol{u} = \boldsymbol{u}_{\mathrm{D}} \quad \text{on } \partial\Omega_{\mathrm{D}}, \tag{4.4}$$

$$\boldsymbol{\sigma}(\boldsymbol{\varepsilon}(\boldsymbol{u}))\boldsymbol{n} = \boldsymbol{g}_{\mathrm{N}} \quad \text{on } \partial\Omega_{\mathrm{N}}, \tag{4.5}$$

together with (4.1) and (4.2). Setting $\boldsymbol{H}^1(\Omega) := H^1(\Omega;\mathbb{R}^d)$, the functional space composed of the kinematically admissible displacements and its tangent space are

$$\boldsymbol{V}_{\mathrm{D}} := \{\boldsymbol{v} \in \boldsymbol{H}^1(\Omega) \mid \boldsymbol{v}_{|\partial\Omega_{\mathrm{D}}} = \boldsymbol{u}_{\mathrm{D}}\}, \tag{4.6}$$

$$\boldsymbol{V}_0 := \{\boldsymbol{v} \in \boldsymbol{H}^1(\Omega) \mid \boldsymbol{v}_{|\partial\Omega_{\mathrm{D}}} = \boldsymbol{0}\}. \tag{4.7}$$

Recall that the $\boldsymbol{H}^1$-norm is defined as $\|\boldsymbol{v}\|_{\boldsymbol{H}^1(\Omega)} := \big(\|\boldsymbol{v}\|^2_{\boldsymbol{L}^2(\Omega)} + \ell_\Omega^2\|\nabla\boldsymbol{v}\|^2_{\mathbb{L}^2(\Omega)}\big)^{\frac{1}{2}}$, where the length scale $\ell_\Omega := \mathrm{diam}\,(\Omega)$ is introduced to be dimensionally consistent. Since $|\partial\Omega_{\mathrm{D}}| > 0$, Korn's inequality implies that there is $C_{\mathrm{K}} > 0$ such that $C_{\mathrm{K}}\|\nabla\boldsymbol{v}\|_{\mathbb{L}^2(\Omega)} \le \|\boldsymbol{\varepsilon}(\boldsymbol{v})\|_{\mathbb{L}^2(\Omega)}$ for all $\boldsymbol{v} \in \boldsymbol{V}_0$ (see, e.g., [108], [122, Theorem 10.2]). Moreover, the Poincaré–Steklov inequality applied componentwise shows that there is $C_{\mathrm{PS}} > 0$ such that $C_{\mathrm{PS}}\|\boldsymbol{v}\|_{\boldsymbol{L}^2(\Omega)} \le \ell_\Omega\|\nabla\boldsymbol{v}\|_{\mathbb{L}^2(\Omega)}$ for all $\boldsymbol{v} \in \boldsymbol{V}_0$. Assuming $\boldsymbol{f} \in$

$\boldsymbol{L}^2(\Omega)$ and $\boldsymbol{g}_\mathrm{N} \in \boldsymbol{L}^2(\partial\Omega_\mathrm{N})$, the weak formulation of the linear elasticity problem is as follows: Seek $\boldsymbol{u} \in \boldsymbol{V}_\mathrm{D}$ such that

$$a(\boldsymbol{u}, \boldsymbol{w}) = \ell(\boldsymbol{w}) := (\boldsymbol{f}, \boldsymbol{w})_{\boldsymbol{L}^2(\Omega)} + (\boldsymbol{g}_\mathrm{N}, \boldsymbol{w})_{\boldsymbol{L}^2(\partial\Omega_\mathrm{N})}, \quad \forall \boldsymbol{w} \in \boldsymbol{V}_0, \tag{4.8}$$

with the bilinear form $a : \boldsymbol{H}^1(\Omega) \times \boldsymbol{H}^1(\Omega) \to \mathbb{R}$ such that

$$a(\boldsymbol{v}, \boldsymbol{w}) := (\boldsymbol{\sigma}(\boldsymbol{\varepsilon}(\boldsymbol{v})), \boldsymbol{\varepsilon}(\boldsymbol{w}))_{\mathbb{L}^2(\Omega)} = 2\mu(\boldsymbol{\varepsilon}(\boldsymbol{v}), \boldsymbol{\varepsilon}(\boldsymbol{w}))_{\mathbb{L}^2(\Omega)} + \lambda(\nabla\cdot\boldsymbol{v}, \nabla\cdot\boldsymbol{w})_{L^2(\Omega)}, \tag{4.9}$$

where we used that $\mathrm{tr}(\boldsymbol{\varepsilon}(\boldsymbol{v})) = \nabla\cdot\boldsymbol{v}$. Simple manipulations show that

$$2\mu\|\boldsymbol{\varepsilon}\|_{\ell^2}^2 + \lambda|\,\mathrm{tr}(\boldsymbol{\varepsilon})|^2 \geq \min(2\mu, 3\kappa)\|\boldsymbol{\varepsilon}\|_{\ell^2}^2, \tag{4.10}$$

where $\|\boldsymbol{\varepsilon}\|_{\ell^2}^2 := \boldsymbol{\varepsilon}{:}\boldsymbol{\varepsilon} = \sum_{1\leq i,j\leq d} |\boldsymbol{\varepsilon}_{ij}|^2$. Combining this bound with the Korn and Poincaré–Steklov inequalities shows that the bilinear form a is coercive on $\boldsymbol{V}_0$. Hence, after lifting the Dirichlet datum, one can show that the model problem (4.8) is well-posed by invoking the Lax–Milgram lemma.

Standard convexity arguments show that the weak solution $\boldsymbol{u} \in \boldsymbol{V}_\mathrm{D}$ to (4.8) is the unique minimizer in $\boldsymbol{V}_\mathrm{D}$ of the energy functional $\mathfrak{E} : \boldsymbol{V}_\mathrm{D} \to \mathbb{R}$ such that

$$\mathfrak{E}(\boldsymbol{v}) := \frac{1}{2}\Big(2\mu\|\boldsymbol{\varepsilon}(\boldsymbol{v})\|_{\mathbb{L}^2(\Omega)}^2 + \lambda\|\nabla\cdot\boldsymbol{v}\|_{L^2(\Omega)}^2\Big) - \ell(\boldsymbol{v}). \tag{4.11}$$

Moreover, the weak formulation (4.8) expresses the principle of virtual work, wherein the test function $\boldsymbol{w}$ plays the role of a virtual displacement.

Remark 4.1 (*Rigid-body motions*) An important fact in continuum mechanics is that the gradient and strain operators have different kernels. In fact, $\nabla\boldsymbol{v} = \boldsymbol{0}$ if and only if there is $\boldsymbol{a} \in \mathbb{R}^d$ such that $\boldsymbol{v} = \boldsymbol{a}$, i.e., the displacement field $\boldsymbol{v}$ represents a translation. Instead, $\boldsymbol{\varepsilon}(\boldsymbol{v}) = \boldsymbol{0}$ if and only if $\boldsymbol{v} \in \boldsymbol{RM}$ where

$$\boldsymbol{RM} := \begin{cases} \mathbb{P}_3^0 + \boldsymbol{x}\times\mathbb{P}_3^0 & \text{if } d = 3; \quad \dim(\boldsymbol{RM}) = 6, \\ \mathbb{P}_2^0 + \boldsymbol{x}^\perp\mathbb{P}_2^0 & \text{if } d = 2; \quad \dim(\boldsymbol{RM}) = 3, \end{cases} \tag{4.12}$$

where $\mathbb{P}_d^0 := \mathbb{P}_d^0(\mathbb{R}^d;\mathbb{R}^d)$, $\boldsymbol{x}$ is the position vector in $\mathbb{R}^d$, $\times$ denotes the cross product in $\mathbb{R}^3$ and $\boldsymbol{x}^\perp = (x, y)^\perp = (-y, x)$ if $d = 2$. Notice that for $d \in \{2, 3\}$, we have $\mathbb{P}_d^0 \subsetneq \boldsymbol{RM} \subsetneq \mathbb{P}_d^1$. Fields in $\boldsymbol{RM}$ are called rigid-body motions (or translation-rotation motions). □

4.1.2 *Finite Deformations and Hyperelasticity*

We are now concerned with finite deformations. We adopt the Lagrangian description so that all the differential operators are taken with respect to the coordinates in the reference configuration. The deformations are measured using the deformation gradient

$$F(u) := I_d + \nabla u, \tag{4.13}$$

taking values in the set $\mathbb{R}_+^{d\times d}$ of $d \times d$ matrices with positive determinant. In the setting of homogeneous hyperelastic materials, the internal efforts in the body are described at any point in Ω by the first Piola–Kirchhoff tensor P which depends (nonlinearly) on the deformation gradient F at that point. The constitutive relation between P and F is derived by postulating a strain energy density $\Psi : \mathbb{R}_+^{d\times d} \to \mathbb{R}$ and setting

$$P(F) := \partial_F \Psi(F). \tag{4.14}$$

We will mainly deal with hyperelastic materials of Neohookean type extended to the compressible range such that

$$\Psi(F) := \frac{\mu}{2}(F{:}F - d) - \mu \ln J + \frac{\lambda}{2}(\ln J)^2, \qquad J := \det(F), \tag{4.15}$$

where μ and λ are material constants. Since $\partial_F J = JF^{-\mathsf{T}}$, (4.15) gives

$$P(F) = \mu(F - F^{-\mathsf{T}}) + \lambda \ln J F^{-\mathsf{T}}. \tag{4.16}$$

In the above setting, the displacement field $u : \Omega \to \mathbb{R}^d$ satisfies the following equations:

$$-\nabla{\cdot}P(F(u)) = f \qquad \text{in } \Omega, \tag{4.17}$$

$$u = u_{\mathrm{D}} \qquad \text{on } \partial\Omega_{\mathrm{D}}, \tag{4.18}$$

$$P(F(u))\,n = g_{\mathrm{N}} \qquad \text{on } \partial\Omega_{\mathrm{N}}, \tag{4.19}$$

together with (4.13) and (4.14), and where we assumed so-called dead external forces f and g_{N} (i.e., independent of the deformed configuration). Defining the energy functional $\mathfrak{E} : V_{\mathrm{D}} \to \mathbb{R}$ such that

$$\mathfrak{E}(v) := \int_\Omega \Psi(F(v))\,\mathrm{d}x - \ell(v), \tag{4.20}$$

with the linear form ℓ defined in (4.8), the static equilibrium problem (4.17)–(4.19) consists of seeking the stationary points of the energy functional $\mathfrak{E}$ which satisfy the following weak form of the Euler–Lagrange equations:

$$0 = D\mathfrak{E}(u)[v] = \int_\Omega P(F(u)) : \nabla v\,\mathrm{d}x - \ell(v), \tag{4.21}$$

for all virtual displacements $\boldsymbol{v} \in \boldsymbol{V}_0$. We assume that the strain energy density function Ψ is polyconvex, so that local minimizers of the energy functional exist (see [12]). We refer the reader to the textbooks [20, 56, 130] for further insight into the physical modeling.

4.2 HHO Methods for Linear Elasticity

The goal of this section is to present and analyze the HHO method to discretize the linear elasticity problem introduced in Sect. 4.1.1. We assume in the whole section that $\lambda \geq 0$.

4.2.1 *Discrete Unknowns, Reconstruction, and Stabilization*

Let $\mathcal{T}$ be a mesh of Ω belonging to a shape-regular mesh sequence (see Sects. 1.2.1 and 2.1.1). We additionally require that every mesh cell $T \in \mathcal{T}$ is star-shaped with respect to every point in a ball with radius uniformly equivalent to h_T; this will allow us to invoke a local Korn inequality. We assume that Ω is a polyhedron so that the mesh covers Ω exactly. Moreover, we assume that every mesh boundary face belongs either to $\partial\Omega_{\mathrm{D}}$ or to $\partial\Omega_{\mathrm{N}}$; the corresponding subsets of $\mathcal{F}^{\partial}$ are denoted by $\mathcal{F}^{\partial}_{\mathrm{D}}$ and $\mathcal{F}^{\partial}_{\mathrm{N}}$. Recall that in HHO methods, the discrete unknowns are polynomials attached to the mesh cells and the mesh faces. In the context of continuum mechanics, both unknowns are vector-valued: the cell unknowns approximate the displacement field in the cell, and the face unknowns approximate its trace on the mesh faces. For brevity, we only consider the equal-order setting for the cell and face unknowns. One important difference with the diffusion model problem is that we now take the polynomial degree $k \geq 1$. The reason for excluding the case $k = 0$ is related to the necessity to control the rigid-body motions in each mesh cell (for a lowest-order nonconforming method, see [25]).

For every mesh cell $T \in \mathcal{T}$, we set

$$\hat{\boldsymbol{V}}^k_T := \mathbb{P}^k_d(T) \times \mathbb{P}^k_{d-1}(\mathcal{F}_T), \qquad \mathbb{P}^k_{d-1}(\mathcal{F}_T) := \bigtimes_{F\in\mathcal{F}_T} \mathbb{P}^k_{d-1}(F), \tag{4.22}$$

with $\mathbb{P}^k_d(T) := \mathbb{P}^k_d(T; \mathbb{R}^d)$ and $\mathbb{P}^k_{d-1}(F) := \mathbb{P}^k_{d-1}(F; \mathbb{R}^d)$; see Fig. 4.1. A generic element in $\hat{\boldsymbol{V}}^k_T$ is denoted by $\hat{\boldsymbol{v}}_T := (\boldsymbol{v}_T, \boldsymbol{v}_{\partial T})$. The HHO space is defined as

$$\hat{\boldsymbol{V}}^k_h := \boldsymbol{V}^k_{\mathcal{T}} \times \boldsymbol{V}^k_{\mathcal{F}}, \qquad \boldsymbol{V}^k_{\mathcal{T}} := \bigtimes_{T\in\mathcal{T}} \mathbb{P}^k_d(T), \qquad \boldsymbol{V}^k_{\mathcal{F}} := \bigtimes_{F\in\mathcal{F}} \mathbb{P}^k_{d-1}(F), \tag{4.23}$$

so that $\dim(\hat{\boldsymbol{V}}^k_h) = d\binom{k+d}{d}\#\mathcal{T} + d\binom{k+d-1}{d-1}\#\mathcal{F}$. A generic element in $\hat{\boldsymbol{V}}^k_h$ is denoted by $\hat{\boldsymbol{v}}_h := (\boldsymbol{v}_{\mathcal{T}}, \boldsymbol{v}_{\mathcal{F}})$ with $\boldsymbol{v}_{\mathcal{T}} := (\boldsymbol{v}_T)_{T\in\mathcal{T}}$ and $\boldsymbol{v}_{\mathcal{F}} := (\boldsymbol{v}_F)_{F\in\mathcal{F}}$, and we localize the

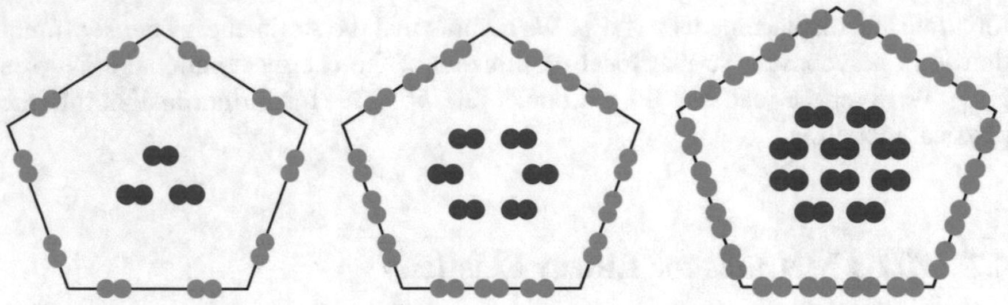

(a) $k = 1$ (b) $k = 2$ (c) $k = 3$

Fig. 4.1 Face (violet) and cell (blue) unknowns in $\hat{V}_T^k$ in a pentagonal cell ($d = 2$) for $k \in \{1, 2, 3\}$ (each dot in a pair represents one basis function associated with one Cartesian component)

components of $\hat{\boldsymbol{v}}_h$ associated with a mesh cell $T \in \mathcal{T}$ and its faces by using the notation $\hat{\boldsymbol{v}}_T := \big(\boldsymbol{v}_T, \boldsymbol{v}_{\partial T} := (\boldsymbol{v}_F)_{F\in\mathcal{F}_T}\big) \in \hat{\boldsymbol{V}}_T^k$. The local HHO reduction operator $\hat{\boldsymbol{I}}_T^k : \boldsymbol{H}^1(T) \to \hat{\boldsymbol{V}}_T^k$ is such that

$$\hat{\boldsymbol{I}}_T^k(\boldsymbol{v}) := (\boldsymbol{\Pi}_T^k(\boldsymbol{v}), \boldsymbol{\Pi}_{\partial T}^k(\boldsymbol{v}_{|\partial T})), \tag{4.24}$$

and its global counterpart $\hat{\boldsymbol{I}}_h^k : \boldsymbol{H}^1(\Omega) \to \hat{\boldsymbol{V}}_h^k$ is $\hat{\boldsymbol{I}}_h^k(\boldsymbol{v}) := (\boldsymbol{\Pi}_{\mathcal{T}}^k(\boldsymbol{v}), \boldsymbol{\Pi}_{\mathcal{F}}^k(\boldsymbol{v}_{|\mathcal{F}}))$, where the L^2-orthogonal projections act componentwise.

The present HHO method for linear elasticity hinges on three local operators: (i) a displacement reconstruction operator, (ii) a divergence reconstruction operator, and (iii) a stabilization operator. The displacement reconstruction operator is based on inverting the Gram matrix associated with the strain tensor. For this reason, we need to specify additionally the rigid-body motions of the reconstructed displacement. For a polynomial degree $l \geq 1$, we set

$$\mathbb{P}_d^l(T)^{\perp} := \{\boldsymbol{q} \in \mathbb{P}_d^l(T) \mid \boldsymbol{\Pi}_T^0(\boldsymbol{q}) = \boldsymbol{0},\ \boldsymbol{\Pi}_T^0(\nabla_{\mathrm{ss}}\boldsymbol{q}) = \boldsymbol{0}\}, \tag{4.25}$$

where $\nabla_{\mathrm{ss}}\boldsymbol{v} := \frac{1}{2}(\nabla\boldsymbol{v} - \nabla\boldsymbol{v}^{\top})$ denotes the skew-symmetric part of the gradient (recall that the $\boldsymbol{L}^2$-orthogonal projection $\boldsymbol{\Pi}_T^0$ returns the mean-value over T). It is easy to see that $\mathbb{P}_d^l(T) = \mathbb{P}_d^l(T)^{\perp} \oplus \boldsymbol{RM}$, where $\boldsymbol{RM}$ is the space of rigid-body motions defined in (4.12). The displacement reconstruction operator $\boldsymbol{U}_T : \hat{\boldsymbol{V}}_T^k \to \mathbb{P}_d^{k+1}(T)$ is defined such that for all $\hat{\boldsymbol{v}}_T \in \hat{\boldsymbol{V}}_T^k$, $\boldsymbol{U}_T(\hat{\boldsymbol{v}}_T) \in \mathbb{P}_d^{k+1}(T)$ is uniquely specified as follows:

$$(\boldsymbol{\varepsilon}(\boldsymbol{U}_T(\hat{\boldsymbol{v}}_T)), \boldsymbol{\varepsilon}(\boldsymbol{q}))_{\boldsymbol{L}^2(T)} = -(\boldsymbol{v}_T, \nabla{\cdot}\boldsymbol{\varepsilon}(\boldsymbol{q}))_{\boldsymbol{L}^2(T)} + (\boldsymbol{v}_{\partial T}, \boldsymbol{\varepsilon}(\boldsymbol{q})\boldsymbol{n}_T)_{\boldsymbol{L}^2(\partial T)}, \tag{4.26}$$

$$\boldsymbol{\Pi}_T^0(\boldsymbol{U}_T(\hat{\boldsymbol{v}}_T)) = \boldsymbol{\Pi}_T^0(\boldsymbol{v}_T), \quad \boldsymbol{\Pi}_T^0(\nabla_{\mathrm{ss}}\boldsymbol{U}_T(\hat{\boldsymbol{v}}_T)) = \frac{1}{|T|}\int_{\partial T} \boldsymbol{v}_{\partial T} \otimes_{\mathrm{ss}} \boldsymbol{n}_T \,\mathrm{d}s, \tag{4.27}$$

where (4.26) holds for all $\boldsymbol{q} \in \mathbb{P}_d^{k+1}(T)^{\perp}$ and $\boldsymbol{a} \otimes_{\mathrm{ss}} \boldsymbol{b} := \frac{1}{2}(\boldsymbol{a} \otimes \boldsymbol{b} - \boldsymbol{b} \otimes \boldsymbol{a})$ for all $\boldsymbol{a}, \boldsymbol{b} \in \mathbb{R}^d$. Proceeding as in the proof of Lemma 1.4 and since $\boldsymbol{\Pi}_T^0(\nabla_{\mathrm{ss}}\boldsymbol{U}_T(\hat{\boldsymbol{I}}_T^k(\boldsymbol{v}))) =$

$\frac{1}{|T|}\int_{\partial T} \boldsymbol{\Pi}^k_{\partial T}(\boldsymbol{v}_{|\partial T}) \otimes_{\rm ss} \boldsymbol{n}_T \, {\rm d}s = \boldsymbol{\Pi}^0_T(\nabla_{\rm ss}\boldsymbol{v})$ (owing to (4.27), (4.24), and integration by parts), one readily establishes the following identity.

Lemma 4.2 (Elliptic projection) *We have $\boldsymbol{U}_T \circ \hat{\boldsymbol{I}}^k_T = \mathcal{E}^{k+1}_T$ where $\mathcal{E}^{k+1}_T : \boldsymbol{H}^1(T) \to \mathbb{P}^{k+1}_d(T)$ is the elliptic strain projection such that $(\boldsymbol{\varepsilon}(\mathcal{E}^{k+1}_T(\boldsymbol{v}) - \boldsymbol{v}), \boldsymbol{\varepsilon}(\boldsymbol{q}))_{L^2(T)} = 0$ for all $\boldsymbol{q} \in \mathbb{P}^{k+1}_d(T)^\perp$, $\boldsymbol{\Pi}^0_T(\mathcal{E}^{k+1}_T(\boldsymbol{v}) - \boldsymbol{v}) = \boldsymbol{0}$, and $\boldsymbol{\Pi}^0_T(\nabla_{ss}(\mathcal{E}^{k+1}_T(\boldsymbol{v}) - \boldsymbol{v})) = \boldsymbol{0}$.*

The divergence reconstruction operator $D_T : \hat{\boldsymbol{V}}^k_T \to \mathbb{P}^k_d(T)$ is defined by solving the following well-posed problem: For all $\hat{\boldsymbol{v}}_T \in \hat{\boldsymbol{V}}^k_T$,

$$(D_T(\hat{\boldsymbol{v}}_T), q)_{L^2(T)} = -(\boldsymbol{v}_T, \nabla q)_{\boldsymbol{L}^2(T)} + (\boldsymbol{v}_{\partial T}, q\boldsymbol{n}_T)_{\boldsymbol{L}^2(\partial T)}, \tag{4.28}$$

for all $q \in \mathbb{P}^k_d(T)$. In practice, the computation of $D_T(\hat{\boldsymbol{v}}_T)$ entails inverting the mass matrix in $\mathbb{P}^k_d(T)$. We have the following important commutation result.

Lemma 4.3 (Commuting with divergence) *The following holds true:*

$$D_T(\hat{\boldsymbol{I}}^k_T(\boldsymbol{v})) = \Pi^k_T(\nabla{\cdot}\boldsymbol{v}), \quad \forall \boldsymbol{v} \in \boldsymbol{H}^1(T). \tag{4.29}$$

Proof Let $q \in \mathbb{P}^k_d(T)$. Since $\nabla q \in \mathbb{P}^k_d(T)$ and $q\boldsymbol{n}_T \in \mathbb{P}^k_{d-1}(\mathcal{F}_T)$, we have

$$\begin{aligned}(D_T(\hat{\boldsymbol{I}}^k_T(\boldsymbol{v})), q)_{L^2(T)} &= -(\boldsymbol{\Pi}^k_T(\boldsymbol{v}), \nabla q)_{\boldsymbol{L}^2(T)} + (\boldsymbol{\Pi}^k_{\partial T}(\boldsymbol{v}_{|\partial T}), q\boldsymbol{n}_T)_{\boldsymbol{L}^2(\partial T)} \\ &= -(\boldsymbol{v}, \nabla q)_{\boldsymbol{L}^2(T)} + (\boldsymbol{v}, q\boldsymbol{n}_T)_{\boldsymbol{L}^2(\partial T)} = (\nabla{\cdot}\boldsymbol{v}, q)_{L^2(T)}.\end{aligned}$$

Since $D_T(\hat{\boldsymbol{I}}^k_T(\boldsymbol{v})) \in \mathbb{P}^k_d(T)$ and q is arbitrary in $\mathbb{P}^k_d(T)$, this proves (4.29). □

Finally, the stabilization operator is inspired from the one devised for the Poisson model problem in Sect. 1.3.2. Here, we define $\boldsymbol{S}_{\partial T} : \hat{\boldsymbol{V}}^k_T \to \mathbb{P}^k_{d-1}(\mathcal{F}_T)$ such that for all $\hat{\boldsymbol{v}}_T \in \hat{\boldsymbol{V}}^k_T$,

$$\boldsymbol{S}_{\partial T}(\hat{\boldsymbol{v}}_T) := \boldsymbol{\Pi}^k_{\partial T}\Big(\boldsymbol{v}_{T|\partial T} - \boldsymbol{v}_{\partial T} + \big((I - \boldsymbol{\Pi}^k_T)\boldsymbol{U}_T(\hat{\boldsymbol{v}}_T)\big)_{|\partial T}\Big), \tag{4.30}$$

where I is the identity operator. Letting $\boldsymbol{\delta}_{\partial T} := \boldsymbol{v}_{T|\partial T} - \boldsymbol{v}_{\partial T}$, we observe that $\boldsymbol{S}_{\partial T}(\hat{\boldsymbol{v}}_T) = \boldsymbol{\Pi}^k_{\partial T}\big(\boldsymbol{\delta}_{\partial T} - ((I - \boldsymbol{\Pi}^k_T)\boldsymbol{U}_T(\boldsymbol{0}, \boldsymbol{\delta}_{\partial T}))_{|\partial T}\big)$.

4.2.2 Discrete Problem, Energy Minimization, and Traction Recovery

The global bilinear form $a_h : \hat{\boldsymbol{V}}^k_h \times \hat{\boldsymbol{V}}^k_h \to \mathbb{R}$ is assembled cellwise as in Sect. 1.4.1 by setting $a_h(\hat{\boldsymbol{v}}_h, \hat{\boldsymbol{w}}_h) := \sum_{T\in\mathcal{T}} a_T(\hat{\boldsymbol{v}}_T, \hat{\boldsymbol{w}}_T)$ where for all $T \in \mathcal{T}$, the local bilinear forms $a_T : \hat{\boldsymbol{V}}^k_T \times \hat{\boldsymbol{V}}^k_T \to \mathbb{R}$ are such that

$$\begin{aligned}a_T(\hat{\boldsymbol{v}}_T, \hat{\boldsymbol{w}}_T) := {}& 2\mu(\boldsymbol{\varepsilon}(\boldsymbol{U}_T(\hat{\boldsymbol{v}}_T)), \boldsymbol{\varepsilon}(\boldsymbol{U}_T(\hat{\boldsymbol{w}}_T)))_{\mathbb{L}^2(T)} + \lambda(D_T(\hat{\boldsymbol{v}}_T), D_T(\hat{\boldsymbol{w}}_T))_{L^2(T)} \\ & + 2\mu h_T^{-1}(\boldsymbol{S}_{\partial T}(\hat{\boldsymbol{v}}_T), \boldsymbol{S}_{\partial T}(\hat{\boldsymbol{w}}_T))_{\boldsymbol{L}^2(\partial T)}.\end{aligned} \tag{4.31}$$

Notice that the stabilization term is weighted by the Lamé parameter μ. To account for the Dirichlet boundary condition, we define the subspaces

$$\boldsymbol{V}^k_{\mathcal{F},\mathrm{D}} := \{\boldsymbol{v}_{\mathcal{F}} \in \boldsymbol{V}^k_{\mathcal{F}} \mid \boldsymbol{v}_F = \boldsymbol{\Pi}^k_F(\boldsymbol{u}_\mathrm{D}),\ \forall F \in \mathcal{F}^\partial_\mathrm{D}\}, \tag{4.32}$$

$$\boldsymbol{V}^k_{\mathcal{F},0} := \{\boldsymbol{v}_{\mathcal{F}} \in \boldsymbol{V}^k_{\mathcal{F}} \mid \boldsymbol{v}_F = \boldsymbol{0},\ \forall F \in \mathcal{F}^\partial_\mathrm{D}\}, \tag{4.33}$$

as well as $\hat{\boldsymbol{V}}^k_{h,\mathrm{D}} := \boldsymbol{V}^k_{\mathcal{T}} \times \boldsymbol{V}^k_{\mathcal{F},\mathrm{D}}$ and $\hat{\boldsymbol{V}}^k_{h,0} := \boldsymbol{V}^k_{\mathcal{T}} \times \boldsymbol{V}^k_{\mathcal{F},0}$. The discrete problem is as follows:

$$\begin{cases} \text{Find } \hat{\boldsymbol{u}}_h \in \hat{\boldsymbol{V}}^k_{h,\mathrm{D}} \text{ such that} \\ a_h(\hat{\boldsymbol{u}}_h, \hat{\boldsymbol{w}}_h) = \ell_h(\hat{\boldsymbol{w}}_h) := (\boldsymbol{f}, \boldsymbol{w}_{\mathcal{T}})_{\boldsymbol{L}^2(\Omega)} + (\boldsymbol{g}_\mathrm{N}, \boldsymbol{w}_{\mathcal{F}})_{\boldsymbol{L}^2(\partial\Omega_\mathrm{N})}, \quad \forall \hat{\boldsymbol{w}}_h \in \hat{\boldsymbol{V}}^k_{h,0}. \end{cases} \tag{4.34}$$

Notice that the cell component of the test function is used against the body force $\boldsymbol{f}$, whereas the face component is used against the traction force $\boldsymbol{g}_\mathrm{N}$. We will see in the next section that the bilinear form a_h is coercive on $\hat{\boldsymbol{V}}^k_{h,0}$ so that the discrete problem (4.34) is well-posed. Moreover, let us define the discrete energy functional $\mathfrak{E}_h : \hat{\boldsymbol{V}}^k_{h,\mathrm{D}} \to \mathbb{R}$ such that

$$\mathfrak{E}_h(\hat{\boldsymbol{v}}_h) := \frac{1}{2}\Big(2\mu\|\boldsymbol{\varepsilon}(\boldsymbol{U}_{\mathcal{T}}(\hat{\boldsymbol{v}}_h))\|^2_{\mathbb{L}^2(\Omega)} + \lambda\|D_{\mathcal{T}}(\hat{\boldsymbol{v}}_h)\|^2_{L^2(\Omega)} + 2\mu s_h(\hat{\boldsymbol{v}}_h, \hat{\boldsymbol{v}}_h)\Big) - \ell_h(\hat{\boldsymbol{v}}_h), \tag{4.35}$$

where the global reconstructions $\boldsymbol{U}_{\mathcal{T}}(\hat{\boldsymbol{v}}_h)$ and $D_{\mathcal{T}}(\hat{\boldsymbol{v}}_h)$ are such that $\boldsymbol{U}_{\mathcal{T}}(\hat{\boldsymbol{v}}_h)_{|T} := \boldsymbol{U}_T(\hat{\boldsymbol{v}}_T)$ and $D_{\mathcal{T}}(\hat{\boldsymbol{v}}_h)_{|T} := D_T(\hat{\boldsymbol{v}}_T)$ for all $T \in \mathcal{T}$, and with the global stabilization bilinear form

$$s_h(\hat{\boldsymbol{v}}_h, \hat{\boldsymbol{w}}_h) := \sum_{T\in\mathcal{T}} h_T^{-1}(\boldsymbol{S}_{\partial T}(\hat{\boldsymbol{v}}_T), \boldsymbol{S}_{\partial T}(\hat{\boldsymbol{w}}_T))_{\boldsymbol{L}^2(\partial T)}. \tag{4.36}$$

Then, the same arguments as in the proof of Proposition 1.8 show that $\hat{\boldsymbol{u}}_h \in \hat{\boldsymbol{V}}^k_{h,\mathrm{D}}$ solves (4.34) if and only if $\hat{\boldsymbol{u}}_h$ minimizes $\mathfrak{E}_h$ in $\hat{\boldsymbol{V}}^k_{h,\mathrm{D}}$.

The algebraic realization of (4.34) leads to a linear system with symmetric positive-definite stiffness matrix having the same block-structure as in (1.39). The right-hand side vector can now have nonzero face components due to the Neumann boundary condition. In any case, a computationally effective way to solve the linear system is again to use static condensation: one eliminates locally all the cell unknowns, solves the global transmission problem coupling all the face unknowns, and finally recovers the cell unknowns by local post-processing. Moreover, the result of Proposition 1.10 on the global transmission problem can be readily extended to the setting of linear elasticity provided the right-hand side of (1.44) is modified to include the contribution of the Neumann boundary condition.

The material of Sect. 1.5.1 on flux recovery can be readily adapted to the present setting leading to the important notion of equilibrated tractions defined on the boundary of the mesh cells and at the Neumann boundary faces. Let $\tilde{\boldsymbol{S}}_{\partial T} : \mathbb{P}^k_{d-1}(\mathcal{F}_T) \to \mathbb{P}^k_{d-1}(\mathcal{F}_T)$ be defined such that

$$\tilde{S}_{\partial T}(\mu) := \Pi^k_{\partial T}\Big(\mu - \big((I - \Pi^k_T)U_T(0,\mu)\big)_{|\partial T}\Big), \tag{4.37}$$

and let $\tilde{S}^*_{\partial T} : \mathbb{P}^k_{d-1}(\mathcal{F}_T) \to \mathbb{P}^k_{d-1}(\mathcal{F}_T)$ be its adjoint such that $(\tilde{S}^*_{\partial T}(\boldsymbol{\lambda}), \boldsymbol{\mu})_{L^2(\partial T)} = (\boldsymbol{\lambda}, \tilde{S}_{\partial T}(\boldsymbol{\mu}))_{L^2(\partial T)}$ for all $\boldsymbol{\lambda}, \boldsymbol{\mu} \in \mathbb{P}^k_{d-1}(\mathcal{F}_T)$. Then, for all $\hat{\boldsymbol{v}}_h \in \hat{\boldsymbol{V}}^k_h$, we can define numerical tractions $\boldsymbol{T}_{\partial T}(\hat{\boldsymbol{v}}_T) \in \mathbb{P}^k_{d-1}(\mathcal{F}_T)$ at the boundary of every mesh cell $T \in \mathcal{T}$ by setting

$$\boldsymbol{T}_{\partial T}(\hat{\boldsymbol{v}}_T) := -\boldsymbol{\sigma}(\boldsymbol{\varepsilon}(\boldsymbol{U}_T(\hat{\boldsymbol{v}}_T)))_{|\partial T}\boldsymbol{n}_T + 2\mu h_T^{-1}(\tilde{S}^*_{\partial T} \circ \tilde{S}_{\partial T})(\boldsymbol{v}_{T|\partial T} - \boldsymbol{v}_{\partial T}). \tag{4.38}$$

A direct adaptation of the proof of Proposition 1.11 establishes the following result.

Proposition 4.4 (Rewriting with tractions) *Let $\hat{\boldsymbol{u}}_h \in \hat{\boldsymbol{V}}^k_{h,D}$ solve (4.34) and let the tractions $\boldsymbol{T}_{\partial T}(\hat{\boldsymbol{u}}_T) \in \mathbb{P}^k_{d-1}(\mathcal{F}_T)$ be defined as in (4.38) for all $T \in \mathcal{T}$. The following holds:*
(i) *Equilibrium at every mesh interface $F = \partial T_- \cap \partial T_+ \cap H_F \in \mathcal{F}^\circ$:*

$$\boldsymbol{T}_{\partial T_-}(\hat{\boldsymbol{u}}_{T_-})_{|F} + \boldsymbol{T}_{\partial T_+}(\hat{\boldsymbol{u}}_{T_+})_{|F} = \boldsymbol{0}, \tag{4.39}$$

and at every Neumann boundary face $F = \partial T_- \cap \partial\Omega_N \cap H_F \in \mathcal{F}^\partial_N$:

$$\boldsymbol{T}_{\partial T_-}(\hat{\boldsymbol{u}}_{T_-})_{|F} + \Pi^k_F(\boldsymbol{g}_N) = \boldsymbol{0}. \tag{4.40}$$

(ii) *Balance with the source term in every mesh cell $T \in \mathcal{T}$: For all $\boldsymbol{q} \in \mathbb{P}^k_d(T)$,*

$$(\boldsymbol{\sigma}(\boldsymbol{\varepsilon}(\boldsymbol{U}_T(\hat{\boldsymbol{u}}_T))), \boldsymbol{\varepsilon}(\boldsymbol{q}))_{L^2(T)} + (\boldsymbol{T}_{\partial T}(\hat{\boldsymbol{u}}_T), \boldsymbol{q})_{L^2(\partial T)} = (\boldsymbol{f}, \boldsymbol{q})_{L^2(T)}. \tag{4.41}$$

(iii) (4.39)–(4.40)–(4.41) *are an equivalent rewriting of* (4.34).

Remark 4.5 (*Literature*) HHO methods for linear elasticity were introduced in [77]. HDG methods for linear elasticity were developed, among others, in [98, 126, 139], weak Galerkin methods in [146], and a hybridizable weakly conforming Galerkin method in [114]. □

4.2.3 Stability and Error Analysis

The stability and error analysis relies on the vector-valued version of the inequalities from Sect. 2.1. In particular, we need the multiplicative trace inequality from Lemma 2.2 and the discrete inverse inequalities from Lemma 2.4. Moreover, in addition to the local Poincaré–Steklov inequality from Lemma 2.1, we need the following local Korn inequality (see [22, Appendix A.1]): There is C_{K} such that for all $T \in \mathcal{T}$,

$$\|\nabla \boldsymbol{v}\|_{L^2(T)} \le C_{\mathrm{K}}\|\boldsymbol{\varepsilon}(\boldsymbol{v})\|_{L^2(T)}, \quad \forall \boldsymbol{v} \in \boldsymbol{H}^1(T)^\perp, \tag{4.42}$$

with $\boldsymbol{H}^1(T)^\perp := \{\boldsymbol{v} \in \boldsymbol{H}^1(T) \mid \boldsymbol{\Pi}_T^0(\boldsymbol{v}) = \boldsymbol{0},\ \boldsymbol{\Pi}_T^0(\nabla_{\mathrm{ss}}\boldsymbol{v}) = \boldsymbol{0}\}$. Combining the local Poincaré–Steklov and Korn inequalities yields $\|\boldsymbol{v}\|_{\boldsymbol{L}^2(T)} \le C_{\mathrm{PS}}C_{\mathrm{K}}h_T\|\boldsymbol{\varepsilon}(\boldsymbol{v})\|_{\mathbb{L}^2(T)}$ for all $\boldsymbol{v} \in \boldsymbol{H}^1(T)^\perp$.

Lemma 4.6 (Stability) *Equip $\hat{\boldsymbol{V}}_T^k$ with the seminorm $|\hat{\boldsymbol{v}}_T|_{\boldsymbol{\varepsilon},T}^2 := \|\boldsymbol{\varepsilon}(\boldsymbol{v}_T)\|_{\mathbb{L}^2(T)}^2 + h_T^{-1}\|\boldsymbol{v}_T - \boldsymbol{v}_{\partial T}\|_{\boldsymbol{L}^2(\partial T)}^2$. Assume $k \ge 1$. There are $0 < \alpha \le \omega < +\infty$ such that for all $T \in \mathcal{T}$ and all $\hat{\boldsymbol{v}}_T \in \hat{\boldsymbol{V}}_T^k$,*

$$\alpha|\hat{\boldsymbol{v}}_T|_{\boldsymbol{\varepsilon},T}^2 \le \|\boldsymbol{\varepsilon}(\boldsymbol{U}_T(\hat{\boldsymbol{v}}_T))\|_{\mathbb{L}^2(T)}^2 + h_T^{-1}\|\boldsymbol{S}_{\partial T}(\hat{\boldsymbol{v}}_T)\|_{\boldsymbol{L}^2(\partial T)}^2 \le \omega|\hat{\boldsymbol{v}}_T|_{\boldsymbol{\varepsilon},T}^2. \tag{4.43}$$

Proof The only difference with the proof of Lemma 2.6 arises in the proof of (2.13) when bounding $(I - \boldsymbol{\Pi}_T^k)(\boldsymbol{U}_T(\hat{\boldsymbol{v}}_T))$. We first notice that there is (a unique) $\boldsymbol{r}_U \in \boldsymbol{RM}$ such that $\boldsymbol{\Pi}_T^0(\boldsymbol{U}_T(\hat{\boldsymbol{v}}_T) - \boldsymbol{r}_U) = \boldsymbol{0}$ and $\boldsymbol{\Pi}_T^0(\nabla_{\mathrm{ss}}(\boldsymbol{U}_T(\hat{\boldsymbol{v}}_T) - \boldsymbol{r}_U)) = \boldsymbol{0}$. Since $(I - \boldsymbol{\Pi}_T^k)(\boldsymbol{r}_U) = \boldsymbol{0}$ (because $k \ge 1$), $\boldsymbol{U}_T(\hat{\boldsymbol{v}}_T) - \boldsymbol{r}_U \in \boldsymbol{H}^1(T)^\perp$, and $\boldsymbol{\varepsilon}(\boldsymbol{r}_U) = \boldsymbol{0}$, we infer that

$$\begin{aligned}
\|(I - \boldsymbol{\Pi}_T^k)(\boldsymbol{U}_T(\hat{\boldsymbol{v}}_T))\|_{\boldsymbol{L}^2(T)} &= \|(I - \boldsymbol{\Pi}_T^k)(\boldsymbol{U}_T(\hat{\boldsymbol{v}}_T) - \boldsymbol{r}_U)\|_{\boldsymbol{L}^2(T)} \\
&\le \|\boldsymbol{U}_T(\hat{\boldsymbol{v}}_T) - \boldsymbol{r}_U\|_{\boldsymbol{L}^2(T)} \\
&\le C_{\mathrm{PS}}C_{\mathrm{K}}h_T\|\boldsymbol{\varepsilon}(\boldsymbol{U}_T(\hat{\boldsymbol{v}}_T) - \boldsymbol{r}_U)\|_{\mathbb{L}^2(T)} \\
&= Ch_T\|\boldsymbol{\varepsilon}(\boldsymbol{U}_T(\hat{\boldsymbol{v}}_T))\|_{\mathbb{L}^2(T)},
\end{aligned}$$

owing to the combined Poincaré–Steklov and Korn inequalities. All the other arguments in the proof are the vector-valued version of those invoked for Lemma 2.6. □

Lemma 4.7 (Coercivity, well-posedness) *The map $\hat{\boldsymbol{V}}_{h,0}^k \ni \hat{\boldsymbol{v}}_h \mapsto \|\hat{\boldsymbol{v}}_h\|_{\boldsymbol{\varepsilon},h} := \big(\sum_{T\in\mathcal{T}} |\hat{\boldsymbol{v}}_T|_{\boldsymbol{\varepsilon},T}^2\big)^{\frac{1}{2}} \in [0, +\infty)$ defines a norm on $\hat{\boldsymbol{V}}_{h,0}^k$. Moreover, the discrete bilinear form a_h satisfies the coercivity property*

$$a_h(\hat{\boldsymbol{v}}_h, \hat{\boldsymbol{v}}_h) \ge 2\mu\|\hat{\boldsymbol{v}}_h\|_{\boldsymbol{\varepsilon},h}^2, \quad \forall \hat{\boldsymbol{v}}_h \in \hat{\boldsymbol{V}}_{h,0}^k, \tag{4.44}$$

and the discrete problem (4.34) *is well-posed.*

Proof The only nontrivial property is the definiteness of the map. Let $\hat{\boldsymbol{v}}_h \in \hat{\boldsymbol{V}}_{h,0}^k$ be such that $\|\hat{\boldsymbol{v}}_h\|_{\boldsymbol{\varepsilon},h} = 0$. Then, for all $T \in \mathcal{T}$, $\boldsymbol{v}_T$ is a rigid displacement whose trace on ∂T is $\boldsymbol{v}_{\partial T}$. Since two rigid displacements that coincide on a mesh face are identical, we infer that $\boldsymbol{v}_{\mathcal{T}}$ is a global rigid displacement, and since $\boldsymbol{v}_F = \boldsymbol{0}$ for all $F \in \mathcal{F}_{\mathrm{D}}^\partial$, we conclude that $\boldsymbol{v}_{\mathcal{T}}$ and $\boldsymbol{v}_{\mathcal{F}}$ are zero. Hence, $\|\cdot\|_{\boldsymbol{\varepsilon},h}$ defines a norm on $\hat{\boldsymbol{V}}_{h,0}^k$. The coercivity property (4.44) follows by summing over the mesh cells the lower bound from Lemma 4.6 and recalling that $\lambda \ge 0$ by assumption. Finally, the well-posedness of (4.34) results from the Lax–Milgram lemma. □

To derive an error estimate, we introduce the consistency error such that

$$\langle \delta_h, \hat{\boldsymbol{w}}_h \rangle := \ell_h(\hat{\boldsymbol{w}}_h) - a_h(\hat{\boldsymbol{I}}_h^k(\boldsymbol{u}), \hat{\boldsymbol{w}}_h), \quad \forall \hat{\boldsymbol{w}}_h \in \hat{\boldsymbol{V}}_{h,0}^k, \tag{4.45}$$

and we bound the dual norm $\|\delta_h\|_* := \sup_{\hat{\boldsymbol{w}}_h \in \hat{\boldsymbol{V}}_{h,0}^k} \frac{|\langle \delta_h, \hat{\boldsymbol{w}}_h \rangle|}{\|\hat{\boldsymbol{w}}_h\|_{\boldsymbol{\varepsilon},h}}$. For all $T \in \mathcal{T}$ and all $\boldsymbol{v} \in \boldsymbol{H}^{1+r}(T)$, $\phi \in H^{1+r}(T)$, $r > \frac{1}{2}$, we define the local (semi)norms

$$|\boldsymbol{v}|_{\sharp,T} := \|\boldsymbol{\varepsilon}(\boldsymbol{v})\|_{\mathbb{L}^2(T)} + h_T^{\frac{1}{2}} \|\boldsymbol{\varepsilon}(\boldsymbol{v})\|_{\mathbb{L}^2(\partial T)}, \quad \|\phi\|_{\dagger,T} := \|\phi\|_{L^2(T)} + h_T^{\frac{1}{2}} \|\phi\|_{L^2(\partial T)}. \tag{4.46}$$

The global counterparts are $|\boldsymbol{v}|_{\sharp,\mathcal{T}} := \big(\sum_{T\in\mathcal{T}} |\boldsymbol{v}|_{\sharp,T}^2\big)^{\frac{1}{2}}$ for all $\boldsymbol{v} \in \boldsymbol{H}^{1+r}(\mathcal{T})$ and $\|\phi\|_{\dagger,\mathcal{T}} := \big(\sum_{T\in\mathcal{T}} \|\phi\|_{\dagger,T}^2\big)^{\frac{1}{2}}$ for all $\phi \in H^{1+r}(\mathcal{T})$. Let $\mathcal{E}_{\mathcal{T}}^{k+1}$ be the global elliptic strain projection such that $\mathcal{E}_{\mathcal{T}}^{k+1}(\boldsymbol{v})_{|T} = \mathcal{E}_T^{k+1}(\boldsymbol{v}_{|T})$ for all $T \in \mathcal{T}$ and all $\boldsymbol{v} \in \boldsymbol{H}^1(\Omega)$.

Lemma 4.8 (Consistency) *Assume that the exact solution satisfies $\boldsymbol{u} \in \boldsymbol{H}^{1+r}(\Omega)$, $r > \frac{1}{2}$. There is C, uniform with respect to μ and λ, such that*

$$\|\delta_h\|_* \le C\big(\mu |\boldsymbol{u} - \mathcal{E}_{\mathcal{T}}^{k+1}(\boldsymbol{u})|_{\sharp,\mathcal{T}} + \lambda \|\nabla\cdot\boldsymbol{u} - \Pi_{\mathcal{T}}^k(\nabla\cdot\boldsymbol{u})\|_{\dagger,\mathcal{T}}\big). \tag{4.47}$$

Proof Since the exact solution satisfies $-\nabla\cdot\boldsymbol{\sigma}(\boldsymbol{\varepsilon}(\boldsymbol{u})) = \boldsymbol{f}$ in Ω and $\boldsymbol{\sigma}(\boldsymbol{\varepsilon}(\boldsymbol{u}))\boldsymbol{n} = \boldsymbol{g}_{\mathrm{N}}$ on $\partial\Omega_{\mathrm{N}}$, integrating by parts and using that the normal component of $\boldsymbol{\sigma}(\boldsymbol{\varepsilon}(\boldsymbol{u}))$ is single-valued at every mesh interface and that $\boldsymbol{w}_F$ vanishes at every Dirichlet boundary face, we infer that

$$\begin{aligned}
\ell(\hat{\boldsymbol{w}}_h) &= \sum_{T\in\mathcal{T}} \big((\boldsymbol{\sigma}(\boldsymbol{\varepsilon}(\boldsymbol{u})), \boldsymbol{\varepsilon}(\boldsymbol{w}_T))_{\mathbb{L}^2(T)} - (\boldsymbol{\sigma}(\boldsymbol{\varepsilon}(\boldsymbol{u}))\boldsymbol{n}_T, \boldsymbol{w}_T)_{\boldsymbol{L}^2(\partial T)}\big) + (\boldsymbol{g}_{\mathrm{N}}, \boldsymbol{w}_{\mathcal{F}})_{\boldsymbol{L}^2(\partial\Omega_{\mathrm{N}})} \\
&= \sum_{T\in\mathcal{T}} \big((\boldsymbol{\sigma}(\boldsymbol{\varepsilon}(\boldsymbol{u})), \boldsymbol{\varepsilon}(\boldsymbol{w}_T))_{\mathbb{L}^2(T)} + (\boldsymbol{\sigma}(\boldsymbol{\varepsilon}(\boldsymbol{u}))\boldsymbol{n}_T, \boldsymbol{w}_{\partial T} - \boldsymbol{w}_T)_{\boldsymbol{L}^2(\partial T)}\big).
\end{aligned}$$

Similar manipulations to the Poisson model problem (see the proof of Lemma 2.8) and the commuting property from Lemma 4.3 give $\langle \delta_h, \hat{\boldsymbol{w}}_h \rangle = -\sum_{T\in\mathcal{T}}(\mathfrak{T}_{1,T} + \mathfrak{T}_{2,T} + \mathfrak{T}_{3,T})$, where

$$\begin{aligned}
\mathfrak{T}_{1,T} &:= 2\mu(\boldsymbol{\varepsilon}(\boldsymbol{u} - \mathcal{E}_T^{k+1}(\boldsymbol{u}))\boldsymbol{n}_T, \boldsymbol{w}_T - \boldsymbol{w}_{\partial T})_{\boldsymbol{L}^2(\partial T)}, \\
\mathfrak{T}_{2,T} &:= \lambda((\nabla\cdot\boldsymbol{u} - \Pi_T^k(\nabla\cdot\boldsymbol{u}))\boldsymbol{n}_T, \boldsymbol{w}_T - \boldsymbol{w}_{\partial T})_{\boldsymbol{L}^2(\partial T)}, \\
\mathfrak{T}_{3,T} &:= 2\mu h_T^{-1}(\boldsymbol{S}_{\partial T}(\hat{\boldsymbol{I}}_T^k(\boldsymbol{u})), \boldsymbol{S}_{\partial T}(\hat{\boldsymbol{w}}_T))_{\boldsymbol{L}^2(\partial T)}.
\end{aligned}$$

The first two terms are bounded by using the Cauchy–Schwarz inequality. The third term is bounded by proceeding as in the proof of Lemma 2.7, except that we additionally invoke the local Korn inequality (4.42). □

As in Lemma 2.9 for the Poisson model problem, stability (Lemma 4.7) and consistency (Lemma 4.8) imply the bound

$$\mu \|\hat{\boldsymbol{e}}_h\|_{\boldsymbol{\varepsilon},h} \le C\big(\mu |\boldsymbol{u} - \mathcal{E}_{\mathcal{T}}^{k+1}(\boldsymbol{u})|_{\sharp,\mathcal{T}} + \lambda \|\nabla\cdot\boldsymbol{u} - \Pi_{\mathcal{T}}^k(\nabla\cdot\boldsymbol{u})\|_{\dagger,\mathcal{T}}\big), \tag{4.48}$$

with the discrete error $\hat{\boldsymbol{e}}_h := \hat{\boldsymbol{u}}_h - \hat{\boldsymbol{I}}_h^k(\boldsymbol{u})$. Using the triangle inequality and the approximation properties of the elliptic strain projection (see [22, Appendix A.2]) leads, as in Theorem 2.10, to the following energy-error estimate. In the spirit of Sect. 2.4, we use the notation $|h^\alpha \boldsymbol{\phi}|_{\boldsymbol{H}^\beta(\mathcal{T})} := \big(\sum_{T\in\mathcal{T}} h_T^{2\alpha} |\boldsymbol{\phi}|^2_{\boldsymbol{H}^\beta(T)}\big)^{\frac{1}{2}}$ for $\boldsymbol{\phi} \in \boldsymbol{H}^\beta(\mathcal{T})$, and we let $\boldsymbol{\varepsilon}_\mathcal{T}$ denote the strain operator applied cellwise (i.e., using the broken gradient).

Theorem 4.9 (Energy-error estimate) *Let $\boldsymbol{u} \in \boldsymbol{V}_D$ be the exact solution and let $\hat{\boldsymbol{u}}_h \in \hat{\boldsymbol{V}}_{h,D}^k$ be the HHO solution solving* (4.34). *Assume that $\boldsymbol{u} \in \boldsymbol{H}^{1+r}(\Omega)$, $r > \frac{1}{2}$. There is C, uniform with respect to μ and λ, such that*

$$\mu\|\boldsymbol{\varepsilon}_\mathcal{T}(\boldsymbol{u} - \boldsymbol{U}_\mathcal{T}(\hat{\boldsymbol{u}}_h))\|_{\boldsymbol{L}^2(\Omega)} \le C\big(\mu|\boldsymbol{u} - \mathcal{E}_\mathcal{T}^{k+1}(\boldsymbol{u})|_{\sharp,\mathcal{T}} + \lambda\|\nabla{\cdot}\boldsymbol{u} - \Pi_\mathcal{T}^k(\nabla{\cdot}\boldsymbol{u})\|_{\dagger,\mathcal{T}}\big). \tag{4.49}$$

Moreover, if $\boldsymbol{u} \in \boldsymbol{H}^{t+1}(\mathcal{T})$ and $\nabla{\cdot}\boldsymbol{u} \in H^t(\mathcal{T})$ for some $t \in (\frac{1}{2}, k+1]$, we have

$$\mu\|\boldsymbol{\varepsilon}_\mathcal{T}(\boldsymbol{u} - \boldsymbol{U}_\mathcal{T}(\hat{\boldsymbol{u}}_h))\|_{\boldsymbol{L}^2(\Omega)} \le C\big(\mu|h^t\boldsymbol{u}|_{\boldsymbol{H}^{t+1}(\mathcal{T})} + \lambda|h^t\nabla{\cdot}\boldsymbol{u}|_{H^t(\mathcal{T})}\big). \tag{4.50}$$

Remark 4.10 (*Quasi-incompressible limit*) The remarkable fact about the error estimates (4.49)–(4.50) is that the right-hand side depends on the second Lamé parameter λ only through the smoothness of $\nabla{\cdot}\boldsymbol{u}$. Furthermore, the incompressible limit (i.e., the Stokes equations) can be treated by introducing a pressure variable attached to the mesh cells [6]. The pressure unknowns, up to the cell mean-value, can be locally eliminated together with the cell velocity unknowns by static condensation. Moreover, as shown in [81], the discretization can be made pressure-robust. □

An improved $\boldsymbol{L}^2$-error estimate can be derived by adapting the arguments presented in Sect. 2.5. We assume the following elliptic regularity property: There are a constant C_{ell} and a regularity pickup index $s \in (\frac{1}{2}, 1]$ such that for all $\boldsymbol{g} \in \boldsymbol{L}^2(\Omega)$, the unique field $\boldsymbol{\zeta}_{\boldsymbol{g}} \in \boldsymbol{V}_0$ such that $a(\boldsymbol{v}, \boldsymbol{\zeta}_{\boldsymbol{g}}) = (\boldsymbol{v}, \boldsymbol{g})_{\boldsymbol{L}^2(\Omega)}$ for all $\boldsymbol{v} \in \boldsymbol{V}_0$ satisfies the regularity estimate $\mu\|\boldsymbol{\zeta}_{\boldsymbol{g}}\|_{\boldsymbol{H}^{1+s}(\Omega)} + \lambda\|\nabla{\cdot}\boldsymbol{\zeta}_{\boldsymbol{g}}\|_{H^s(\Omega)} \le C_{\text{ell}}\ell_\Omega^2\|\boldsymbol{g}\|_{\boldsymbol{L}^2(\Omega)}$. Then, proceeding as in the proof of Lemma 2.11 (see also [77] for the original arguments) leads to the following discrete $\boldsymbol{L}^2$-error estimate: There is C, uniform with respect to μ and λ, such that

$$\begin{aligned}\mu\|\boldsymbol{e}_\mathcal{T}\|_\Omega \le C\ell_\Omega^{1-s}h^s\Big(&\mu|\boldsymbol{u} - \mathcal{E}_\mathcal{T}^{k+1}(\boldsymbol{u})|_{\sharp,\mathcal{T}} + \max(\mu,\lambda)\|\nabla{\cdot}\boldsymbol{u} - \Pi_\mathcal{T}^k(\nabla{\cdot}\boldsymbol{u})\|_{\dagger,\mathcal{T}} \\ &+ h\|\boldsymbol{f} - \boldsymbol{\Pi}_\mathcal{T}^k(\boldsymbol{f})\|_{\boldsymbol{L}^2(\Omega)} + h^{\frac{1}{2}}\|\boldsymbol{g}_\mathrm{N} - \boldsymbol{\Pi}_{\mathcal{F}_\mathrm{N}^\partial}^k(\boldsymbol{g}_\mathrm{N})\|_{\boldsymbol{L}^2(\partial\Omega_\mathrm{N})}\Big).\end{aligned} \tag{4.51}$$

4.3 HHO Methods for Hyperelasticity

The goal of this section is to present and analyze two HHO methods to discretize the hyperelasticity problem introduced in Sect. 4.1.2. Following the ideas outlined in

Sect. 3.1, we consider (i) a stabilized HHO method reconstructing the deformation gradient in $\mathbb{P}_d^k(T;\mathbb{R}^{d\times d})$ and (ii) an unstabilized method reconstructing the deformation gradient in a larger polynomial space built using Raviart–Thomas polynomials. For both methods, the discrete setting is the same as the one considered for the linear elasticity problem (see Sect. 4.2.1): the mesh $\mathcal{T}$ satisfies the assumptions stated therein, the discrete unknowns belong to the local space $\hat{\boldsymbol{V}}_T^k := \mathbf{P}_d^k(T) \times \mathbf{P}_{d-1}^k(\mathcal{F}_T)$ defined in (4.22) for every mesh cell $T \in \mathcal{T}$ and a polynomial degree $k \geq 1$, and the global HHO space is the space $\hat{\boldsymbol{V}}_h^k := \boldsymbol{V}_{\mathcal{T}}^k \times \boldsymbol{V}_{\mathcal{F}}^k$ defined in (4.23) together with the subspaces $\hat{\boldsymbol{V}}_{h,\mathrm{D}}^k := \boldsymbol{V}_{\mathcal{T}}^k \times \boldsymbol{V}_{\mathcal{F},\mathrm{D}}^k$ and $\hat{\boldsymbol{V}}_{h,0}^k := \boldsymbol{V}_{\mathcal{T}}^k \times \boldsymbol{V}_{\mathcal{F},0}^k$ related to the enforcement of the Dirichlet boundary condition (see (4.32)–(4.33)).

4.3.1 The Stabilized HHO Method

The local gradient reconstruction operator $\boldsymbol{G}_T : \hat{\boldsymbol{V}}_T^k \to \mathbb{P}_d^k(T;\mathbb{R}^{d\times d})$ is such that for all $\hat{\boldsymbol{v}}_T \in \hat{\boldsymbol{V}}_T^k$, $\boldsymbol{G}_T(\hat{\boldsymbol{v}}_T) \in \mathbb{P}_d^k(T;\mathbb{R}^{d\times d})$ is uniquely determined by the equations

$$(\boldsymbol{G}_T(\hat{\boldsymbol{v}}_T), \boldsymbol{q})_{L^2(T)} = -(\boldsymbol{v}_T, \nabla{\cdot}\boldsymbol{q})_{L^2(T)} + (\boldsymbol{v}_{\partial T}, \boldsymbol{q}\boldsymbol{n}_T)_{L^2(\partial T)}, \quad \forall \boldsymbol{q} \in \mathbb{P}_d^k(T;\mathbb{R}^{d\times d}). \tag{4.52}$$

To compute $\boldsymbol{G}_T(\hat{\boldsymbol{v}}_T)$, it suffices to invert the mass matrix associated with the scalar-valued polynomial space $\mathbb{P}_d^k(T)$ since only the right-hand side changes when computing each entry of the tensor $\boldsymbol{G}_T(\hat{\boldsymbol{v}}_T)$. Notice in passing that

$$\mathrm{tr}(\boldsymbol{G}_T(\hat{\boldsymbol{v}}_T)) = D_T(\hat{\boldsymbol{v}}_T), \quad \forall \hat{\boldsymbol{v}}_T \in \hat{\boldsymbol{V}}_T^k, \tag{4.53}$$

where D_T is the divergence reconstruction operator defined in (4.28) (take in (4.52) $\boldsymbol{q} := q\boldsymbol{I}_d$ with q arbitrary in $\mathbb{P}_d^k(T)$). Moreover, proceeding as in the proof of Lemma 3.1(ii), one readily verifies that

$$\boldsymbol{G}_T(\hat{\boldsymbol{I}}_T^k(\boldsymbol{v})) = \boldsymbol{\Pi}_T^k(\nabla \boldsymbol{v}), \tag{4.54}$$

for all $\boldsymbol{v} \in \boldsymbol{H}^1(T)$, where $\boldsymbol{\Pi}_T^k$ is the $\boldsymbol{L}^2$-orthogonal projection onto $\mathbb{P}_d^k(T;\mathbb{R}^{d\times d})$ and the local reduction operator $\hat{\boldsymbol{I}}_T^k : \boldsymbol{H}^1(T) \to \hat{\boldsymbol{V}}_T^k$ is defined in (4.24). Finally, the stabilization operator $\boldsymbol{S}_{\partial T} : \hat{\boldsymbol{V}}_T^k \to \mathbf{P}_{d-1}^k(\mathcal{F}_T)$ is defined in (4.30) as for linear elasticity (another possibility is to define it as the vector-valued version of the one used for the Poisson model problem in (1.20)). Adapting the arguments in the proof of Lemma 2.6 leads to the following result.

Lemma 4.11 (Stability) *Equip $\hat{\boldsymbol{V}}_T^k$ with the seminorm $|\hat{\boldsymbol{v}}_T|^2_{\hat{\boldsymbol{V}}_T^k} := \|\nabla \boldsymbol{v}_T\|^2_{\boldsymbol{L}^2(T)} + h_T^{-1}\|\boldsymbol{v}_T - \boldsymbol{v}_{\partial T}\|^2_{\boldsymbol{L}^2(\partial T)}$. There are $0 < \alpha \leq \omega < +\infty$ such that for all $T \in \mathcal{T}$ and all $\hat{\boldsymbol{v}}_T \in \hat{\boldsymbol{V}}_T^k$,*

$$\alpha |\hat{\boldsymbol{v}}_T|^2_{\hat{\boldsymbol{V}}^k_T} \leq \|\boldsymbol{G}_T(\hat{\boldsymbol{v}}_T))\|^2_{\boldsymbol{L}^2(T)} + h_T^{-1}\|\boldsymbol{S}_{\partial T}(\hat{\boldsymbol{v}}_T)\|^2_{\boldsymbol{L}^2(\partial T)} \leq \omega |\hat{\boldsymbol{v}}_T|^2_{\hat{\boldsymbol{V}}^k_T}. \tag{4.55}$$

For all $\hat{\boldsymbol{v}}_T \in \hat{\boldsymbol{V}}^k_T$, we reconstruct the deformation gradient in every mesh cell $T \in \mathcal{T}$ as

$$\boldsymbol{F}_T(\hat{\boldsymbol{v}}_T) := \boldsymbol{I}_d + \boldsymbol{G}_T(\hat{\boldsymbol{v}}_T). \tag{4.56}$$

For all $\hat{\boldsymbol{v}}_h \in \hat{\boldsymbol{V}}^k_h$, we define the global reconstructions $\boldsymbol{F}_{\mathcal{T}}(\hat{\boldsymbol{v}}_h)$ and $\boldsymbol{G}_{\mathcal{T}}(\hat{\boldsymbol{v}}_h)$ such that

$$\boldsymbol{F}_{\mathcal{T}}(\hat{\boldsymbol{v}}_h)_{|T} := \boldsymbol{F}_T(\hat{\boldsymbol{v}}_T), \quad \boldsymbol{G}_{\mathcal{T}}(\hat{\boldsymbol{v}}_h)_{|T} := \boldsymbol{G}_T(\hat{\boldsymbol{v}}_T), \quad \forall T \in \mathcal{T}, \tag{4.57}$$

so that $\boldsymbol{F}_{\mathcal{T}}(\hat{\boldsymbol{v}}_h) = \boldsymbol{I}_d + \boldsymbol{G}_{\mathcal{T}}(\hat{\boldsymbol{v}}_h)$. Recalling the linear form $\ell_h(\hat{\boldsymbol{v}}_h) := (\boldsymbol{f}, \boldsymbol{v}_{\mathcal{T}})_{\boldsymbol{L}^2(\Omega)} + (\boldsymbol{g}_{\mathrm{N}}, \boldsymbol{v}_{\mathcal{F}})_{\boldsymbol{L}^2(\partial\Omega_{\mathrm{N}})}$ from (4.34) and the stabilization bilinear form s_h from (4.36), we define the discrete energy functional $\mathfrak{E}_h : \hat{\boldsymbol{V}}^k_{h,\mathrm{D}} \to \mathbb{R}$ such that (compare with (4.20))

$$\mathfrak{E}_h(\hat{\boldsymbol{v}}_h) := \int_\Omega \Psi(\boldsymbol{F}_{\mathcal{T}}(\hat{\boldsymbol{v}}_h))\,\mathrm{d}\boldsymbol{x} + \beta_0 \mu s_h(\hat{\boldsymbol{v}}_h, \hat{\boldsymbol{v}}_h) - \ell_h(\hat{\boldsymbol{v}}_h), \tag{4.58}$$

with a non-dimensional positive weight $\beta_0 > 0$. For linear elasticity, a simple choice (considered in (4.35)) is $\beta_0 = 1$; for finite deformations of hyperelastic materials, the choice of β_0 is further discussed in Remark 4.17. The discrete problem consists in seeking the stationary points in $\hat{\boldsymbol{V}}^k_{h,\mathrm{D}}$ of the discrete energy functional $\mathfrak{E}_h$: Find $\hat{\boldsymbol{u}}_h \in \hat{\boldsymbol{V}}^k_{h,\mathrm{D}}$ such that (compare with (4.21))

$$(\boldsymbol{P}(\boldsymbol{F}_{\mathcal{T}}(\hat{\boldsymbol{u}}_h)), \boldsymbol{G}_{\mathcal{T}}(\hat{\boldsymbol{w}}_h))_{\boldsymbol{L}^2(\Omega)} + 2\beta_0 \mu s_h(\hat{\boldsymbol{u}}_h, \hat{\boldsymbol{w}}_h) = \ell_h(\hat{\boldsymbol{w}}_h), \quad \forall \hat{\boldsymbol{w}}_h \in \hat{\boldsymbol{V}}^k_{h,0}. \tag{4.59}$$

As for the linear elasticity problem, the discrete problem (4.59) can be reformulated in terms of equilibrated tractions. Let $\tilde{\boldsymbol{S}}_{\partial T} : \mathbb{P}^k_{d-1}(\mathcal{F}_T) \to \mathbb{P}^k_{d-1}(\mathcal{F}_T)$ be defined in (4.37) and let $\tilde{\boldsymbol{S}}^*_{\partial T} : \mathbb{P}^k_{d-1}(\mathcal{F}_T) \to \mathbb{P}^k_{d-1}(\mathcal{F}_T)$ be its adjoint. For all $\hat{\boldsymbol{v}}_h \in \hat{\boldsymbol{V}}^k_h$, we define numerical tractions $\boldsymbol{T}_{\partial T}(\hat{\boldsymbol{v}}_T) \in \mathbb{P}^k_{d-1}(\mathcal{F}_T)$ at the boundary of every mesh cell $T \in \mathcal{T}$ by setting

$$\boldsymbol{T}_{\partial T}(\hat{\boldsymbol{v}}_T) := -\boldsymbol{\Pi}^k_T(\boldsymbol{P}(\boldsymbol{F}_T(\hat{\boldsymbol{v}}_T)))_{|\partial T}\boldsymbol{n}_T + 2\beta_0 \mu h_T^{-1}(\tilde{\boldsymbol{S}}^*_{\partial T} \circ \tilde{\boldsymbol{S}}_{\partial T})(\boldsymbol{v}_{T|\partial T} - \boldsymbol{v}_{\partial T}). \tag{4.60}$$

A direct adaptation of the proof of Proposition 1.11 establishes the following result.

Proposition 4.12 (Rewriting with tractions) *Let $\hat{\boldsymbol{u}}_h \in \hat{\boldsymbol{V}}^k_{h,D}$ solve (4.59) and let the tractions $\boldsymbol{T}_{\partial T}(\hat{\boldsymbol{u}}_T) \in \mathbb{P}^k_{d-1}(\mathcal{F}_T)$ be defined as in (4.60) for all $T \in \mathcal{T}$. The following holds:*
(i) *Equilibrium at every mesh interface $F = \partial T_- \cap \partial T_+ \cap H_F \in \mathcal{F}^\circ$ and at every Neumann boundary face $F = \partial T_- \cap \partial\Omega_N \cap H_F \in \mathcal{F}^\partial_N$: (4.39) and (4.40) hold true.*
(ii) *Balance with the source term in every mesh cell $T \in \mathcal{T}$: For all $\boldsymbol{q} \in \mathbb{P}^k_d(T)$,*

$$(\boldsymbol{P}(\boldsymbol{F}_T(\hat{\boldsymbol{u}}_T)), \nabla \boldsymbol{q})_{\boldsymbol{L}^2(T)} + (\boldsymbol{T}_{\partial T}(\hat{\boldsymbol{u}}_T), \boldsymbol{q})_{\boldsymbol{L}^2(\partial T)} = (\boldsymbol{f}, \boldsymbol{q})_{\boldsymbol{L}^2(T)}. \tag{4.61}$$

(iii) *The above identities are an equivalent rewriting of* (4.59) *that fully characterizes any HHO solution* $\hat{\boldsymbol{u}}_h \in \hat{\boldsymbol{V}}^k_{h,D}$.

Remark 4.13 (*Literature*) HHO methods for hyperelastic materials undergoing finite deformations were introduced in [1], see also [26] for nonlinear elasticity and small deformations. HDG methods for nonlinear elasticity were developed in [111, 126, 138], discontinuous Galerkin methods in [129, 143, 144], gradient schemes in [86], virtual element methods in [50, 151], and a (low-order) hybrid dG method with conforming traces in [153]. □

4.3.2 The Unstabilized HHO Method

In nonlinear elasticity, the use of stabilization can lead to numerical difficulties since it is not clear beforehand how large the stabilization parameter ought to be; see [50, 143] for related discussions. Moreover, [111, Sect. 4] presents an example where spurious solutions can appear in an HDG discretization if the stabilization parameter is not large enough. Motivated by these observations, we present in this section the unstabilized HHO method devised in [1]. We assume for simplicity that the mesh is simplicial.

Let $T \in \mathcal{T}$ and let $k \geq 1$. Let us set

$$\mathbb{RT}^k_d(T;\mathbb{R}^{d\times d}) := \mathbb{P}^k_d(T;\mathbb{R}^{d\times d}) \oplus (\tilde{\mathbb{P}}^k_d(T;\mathbb{R}^d) \otimes \boldsymbol{x}), \tag{4.62}$$

where $\tilde{\mathbb{P}}^k_d(T;\mathbb{R}^d)$ is the space composed of the restriction to T of $\mathbb{R}^d$-valued d-variate homogeneous polynomials of degree k. The local gradient reconstruction operator $\boldsymbol{G}^{\mathrm{RT}}_T : \hat{\boldsymbol{V}}^k_T \to \mathbb{RT}^k_d(T;\mathbb{R}^{d\times d})$ is such that for all $\hat{\boldsymbol{v}}_T \in \hat{\boldsymbol{V}}^k_T$, $\boldsymbol{G}^{\mathrm{RT}}_T(\hat{\boldsymbol{v}}_T) \in \mathbb{RT}^k_d(T;\mathbb{R}^{d\times d})$ is uniquely determined by the equations

$$(\boldsymbol{G}^{\mathrm{RT}}_T(\hat{\boldsymbol{v}}_T), \boldsymbol{q})_{\boldsymbol{L}^2(T)} = -(\boldsymbol{v}_T, \nabla{\cdot}\boldsymbol{q})_{\boldsymbol{L}^2(T)} + (\boldsymbol{v}_{\partial T}, \boldsymbol{q}\boldsymbol{n}_T)_{\boldsymbol{L}^2(\partial T)}, \quad \forall \boldsymbol{q} \in \mathbb{RT}^k_d(T;\mathbb{R}^{d\times d}). \tag{4.63}$$

In practice, the lines of $\boldsymbol{G}^{\mathrm{RT}}_T(\hat{\boldsymbol{v}}_T)$ can be computed separately by inverting the mass matrix associated with the space $\mathbb{RT}^k_d(T;\mathbb{R}^d)$. Notice that the size of the linear system resulting from (4.63) is larger than the one resulting from (4.52); the respective sizes are $d\binom{k+d}{d} + \binom{k+d-1}{d-1}$ versus $\binom{k+d}{d}$, e.g., 15 versus 4 for $d = 3, k = 1$ and 36 versus 10 for $d = 3, k = 2$. Adapting the arguments of the proof of Lemma 3.2(ii) to the tensor-valued case shows that

$$\boldsymbol{G}^{\mathrm{RT}}_T(\hat{\boldsymbol{I}}^k_T(\boldsymbol{v})) = \boldsymbol{\Pi}^k_T(\nabla\boldsymbol{v}), \tag{4.64}$$

for all $\boldsymbol{v} \in \boldsymbol{H}^1(T)$, where $\boldsymbol{\Pi}^k_T$ is the $\boldsymbol{L}^2$-orthogonal projection onto $\mathbb{P}^k_d(T;\mathbb{R}^{d\times d})$ and the local reduction operator $\hat{\boldsymbol{I}}^k_T : \boldsymbol{H}^1(T) \to \hat{\boldsymbol{V}}^k_T$ is defined in (4.24). Adapting the arguments of the proof of Lemma 3.2(iii) for the lower bound and proceeding as usual for the upper bound leads to the following result.

Lemma 4.14 (Stability) *Recall the seminorm* $|\cdot|_{\hat{V}_T^k}$ *from Lemma 4.11. There are* $0 < \alpha \le \omega < +\infty$ *such that* $\alpha|\hat{\boldsymbol{v}}_T|^2_{\hat{V}_T^k} \le \|\boldsymbol{G}_T(\hat{\boldsymbol{v}}_T))\|^2_{\boldsymbol{L}^2(T)} \le \omega|\hat{\boldsymbol{v}}_T|^2_{\hat{V}_T^k}$ *for all* $T \in \mathcal{T}$ *and all* $\hat{\boldsymbol{v}}_T \in \hat{\boldsymbol{V}}_T^k$.

For all $\hat{\boldsymbol{v}}_T \in \hat{\boldsymbol{V}}_T^k$, we now reconstruct the deformation gradient in every mesh cell $T \in \mathcal{T}$ as

$$\boldsymbol{F}_T^{\mathrm{RT}}(\hat{\boldsymbol{v}}_T) := \boldsymbol{I}_d + \boldsymbol{G}_T^{\mathrm{RT}}(\hat{\boldsymbol{v}}_T). \tag{4.65}$$

For all $\hat{\boldsymbol{v}}_h \in \hat{\boldsymbol{V}}_h^k$, we define the global reconstructions $\boldsymbol{F}_{\mathcal{T}}^{\mathrm{RT}}(\hat{\boldsymbol{v}}_h)$ and $\boldsymbol{G}_{\mathcal{T}}^{\mathrm{RT}}(\hat{\boldsymbol{v}}_h)$ such that

$$\boldsymbol{F}_{\mathcal{T}}^{\mathrm{RT}}(\hat{\boldsymbol{v}}_h)_{|T} := \boldsymbol{F}_T^{\mathrm{RT}}(\hat{\boldsymbol{v}}_T), \quad \boldsymbol{G}_{\mathcal{T}}^{\mathrm{RT}}(\hat{\boldsymbol{v}}_h)_{|T} := \boldsymbol{G}_T^{\mathrm{RT}}(\hat{\boldsymbol{v}}_T), \quad \forall T \in \mathcal{T}, \tag{4.66}$$

so that $\boldsymbol{F}_{\mathcal{T}}^{\mathrm{RT}}(\hat{\boldsymbol{v}}_h) = \boldsymbol{I}_d + \boldsymbol{G}_{\mathcal{T}}^{\mathrm{RT}}(\hat{\boldsymbol{v}}_h)$. Recalling the linear form $\ell_h(\hat{\boldsymbol{v}}_h) := (\boldsymbol{f}, \boldsymbol{v}_{\mathcal{T}})_{\boldsymbol{L}^2(\Omega)} + (\boldsymbol{g}_{\mathrm{N}}, \boldsymbol{v}_{\mathcal{F}})_{\boldsymbol{L}^2(\partial\Omega_{\mathrm{N}})}$ (see (4.34)), we define the discrete energy functional $\mathfrak{E}_h : \hat{\boldsymbol{V}}_{h,\mathrm{D}}^k \to \mathbb{R}$ such that (compare with (4.58); we use the same notation for simplicity)

$$\mathfrak{E}_h(\hat{\boldsymbol{v}}_h) := \int_\Omega \Psi(\boldsymbol{F}_{\mathcal{T}}^{\mathrm{RT}}(\hat{\boldsymbol{v}}_h))\,\mathrm{d}\boldsymbol{x} - \ell_h(\hat{\boldsymbol{v}}_h). \tag{4.67}$$

The discrete problem consists in seeking the stationary points in $\hat{\boldsymbol{V}}_{h,\mathrm{D}}^k$ of the discrete energy functional $\mathfrak{E}_h$: Find $\hat{\boldsymbol{u}}_h \in \hat{\boldsymbol{V}}_{h,\mathrm{D}}^k$ such that (compare with (4.21))

$$(\boldsymbol{P}(\boldsymbol{F}_{\mathcal{T}}^{\mathrm{RT}}(\hat{\boldsymbol{u}}_h)), \boldsymbol{G}_{\mathcal{T}}^{\mathrm{RT}}(\hat{\boldsymbol{w}}_h))_{\boldsymbol{L}^2(\Omega)} = \ell_h(\hat{\boldsymbol{w}}_h), \quad \forall \hat{\boldsymbol{w}}_h \in \hat{\boldsymbol{V}}_{h,0}^k. \tag{4.68}$$

The discrete problem (4.68) can be reformulated in terms of equilibrated tractions. For all $\hat{\boldsymbol{v}}_h \in \hat{\boldsymbol{V}}_h^k$, we can define numerical tractions $\boldsymbol{T}_{\partial T}(\hat{\boldsymbol{v}}_T) \in \mathbb{P}_{d-1}^k(\mathcal{F}_T)$ at the boundary of every mesh cell $T \in \mathcal{T}$ by setting

$$\boldsymbol{T}_{\partial T}(\hat{\boldsymbol{v}}_T) := -\boldsymbol{\Pi}_T^{\mathrm{RT}}(\boldsymbol{P}(\boldsymbol{F}_T(\hat{\boldsymbol{v}}_T)))_{|\partial T}\boldsymbol{n}_T, \tag{4.69}$$

where $\boldsymbol{\Pi}_T^{\mathrm{RT}}$ denotes the $\boldsymbol{L}^2$-orthogonal projection onto $\mathbb{RT}_d^k(T; \mathbb{R}^{d\times d})$. A direct adaptation of the proof of Proposition 1.11 establishes the following result.

Proposition 4.15 (Rewriting with tractions) *Let* $\hat{\boldsymbol{u}}_h \in \hat{\boldsymbol{V}}_{h,D}^k$ *solve* (4.68) *and let the tractions* $\boldsymbol{T}_{\partial T}(\hat{\boldsymbol{u}}_T) \in \mathbb{P}_{d-1}^k(\mathcal{F}_T)$ *be defined as in* (4.69) *for all* $T \in \mathcal{T}$. *The following holds:*

(i) *Equilibrium at every mesh interface* $F = \partial T_- \cap \partial T_+ \cap H_F \in \mathcal{F}^\circ$ *and at every Neumann boundary face* $F = \partial T_- \cap \partial\Omega_N \cap H_F \in \mathcal{F}_N^\partial$*:* (4.39) *and* (4.40) *hold true.*
(ii) *Balance with the source term in every mesh cell* $T \in \mathcal{T}$*: For all* $\boldsymbol{q} \in \mathbb{P}_d^k(T)$*,*

$$(\boldsymbol{P}(\boldsymbol{F}_T(\hat{\boldsymbol{u}}_T)), \nabla \boldsymbol{q})_{\boldsymbol{L}^2(T)} + (\boldsymbol{T}_{\partial T}(\hat{\boldsymbol{u}}_T), \boldsymbol{q})_{\boldsymbol{L}^2(\partial T)} = (\boldsymbol{f}, \boldsymbol{q})_{\boldsymbol{L}^2(T)}. \tag{4.70}$$

(iii) *The above identities are an equivalent rewriting of* (4.68).

Remark 4.16 (*Divergence*) Notice that the identity (4.53) no longer holds if the gradient is reconstructed using Raviart–Thomas polynomials. Instead, one only has $\Pi_T^k(\mathrm{tr}(\boldsymbol{G}_T(\hat{\boldsymbol{v}}_T))) = D_T(\hat{\boldsymbol{v}}_T)$ for all $\hat{\boldsymbol{v}}_T \in \hat{\boldsymbol{V}}_T^k$, indicating that a high-order perturbation may hamper robustness in the quasi-incompressible limit. So far, robustness was observed in the numerical experiments. □

4.3.3 Nonlinear Solver and Static Condensation

The nonlinear problems (4.59) and (4.68) can be solved by using Newton's method. This requires evaluating the fourth-order elastic modulus $\mathbb{A}(\boldsymbol{F}) := \partial^2_{\boldsymbol{FF}}\Psi(\boldsymbol{F})$. In particular, for Neohookean materials (see (4.15)), we have

$$\mathbb{A}(\boldsymbol{F}) = \mu(\boldsymbol{I} \overline{\otimes} \boldsymbol{I} + \boldsymbol{F}^{-\mathsf{T}} \underline{\otimes} \boldsymbol{F}^{-1}) - \lambda \ln J \boldsymbol{F}^{-\mathsf{T}} \underline{\otimes} \boldsymbol{F}^{-1} + \lambda \boldsymbol{F}^{-\mathsf{T}} \otimes \boldsymbol{F}^{-\mathsf{T}}, \tag{4.71}$$

with $(\boldsymbol{a} \otimes \boldsymbol{b})_{ijkl} := \boldsymbol{a}_{ij}\boldsymbol{b}_{kl}$, $(\boldsymbol{a} \underline{\otimes} \boldsymbol{b})_{ijkl} := \boldsymbol{a}_{il}\boldsymbol{b}_{jk}$, and $(\boldsymbol{a} \overline{\otimes} \boldsymbol{b})_{ijkl} := \boldsymbol{a}_{ik}\boldsymbol{b}_{jl}$, for all $1 \leq i, j, k, l \leq d$. Let $i \geq 0$ be the index of the Newton's iteration. Given an initial discrete displacement $\hat{\boldsymbol{u}}_h^0 \in \hat{\boldsymbol{V}}_{h,\mathrm{D}}^k$, one computes at each Newton's iteration the incremental displacement $\delta\hat{\boldsymbol{u}}_h^i \in \hat{\boldsymbol{V}}_{h,0}^k$ and updates the discrete displacement as $\hat{\boldsymbol{u}}_h^{i+1} = \hat{\boldsymbol{u}}_h^i + \delta\hat{\boldsymbol{u}}_h^i$. The linear system of equations to be solved is

$$(\mathbb{A}(\boldsymbol{F}_{\mathcal{T}}^*(\hat{\boldsymbol{u}}_h^i)) : \boldsymbol{G}_{\mathcal{T}}^*(\delta\hat{\boldsymbol{u}}_h^i), \boldsymbol{G}_{\mathcal{T}}^*(\hat{\boldsymbol{w}}_h))_{\boldsymbol{L}^2(\Omega)} + 2\beta_0\mu s_h(\delta\hat{\boldsymbol{u}}_h^i, \hat{\boldsymbol{w}}_h) = -R_h^i(\hat{\boldsymbol{w}}_h), \tag{4.72}$$

for all $\hat{\boldsymbol{w}}_h \in \hat{\boldsymbol{V}}_{h,0}^k$, with the residual term

$$R_h^i(\hat{\boldsymbol{w}}_h) := (\boldsymbol{P}(\boldsymbol{F}_{\mathcal{T}}^*(\hat{\boldsymbol{u}}_h^i)), \boldsymbol{G}_{\mathcal{T}}^*(\hat{\boldsymbol{w}}_h))_{\boldsymbol{L}^2(\Omega)} + 2\beta_0\mu s_h(\hat{\boldsymbol{u}}_h^i, \hat{\boldsymbol{w}}_h) - \ell_h(\hat{\boldsymbol{w}}_h), \tag{4.73}$$

where $\beta_0 > 0$, $\boldsymbol{F}_{\mathcal{T}}^* = \boldsymbol{F}_{\mathcal{T}}$, and $\boldsymbol{G}_{\mathcal{T}}^* = \boldsymbol{G}_{\mathcal{T}}$ (see (4.57)) in the stabilized case and $\beta_0 = 0$, $\boldsymbol{F}_{\mathcal{T}}^* = \boldsymbol{F}_{\mathcal{T}}^{\mathrm{RT}}$, and $\boldsymbol{G}_{\mathcal{T}}^* = \boldsymbol{G}_{\mathcal{T}}^{\mathrm{RT}}$ (see (4.66)) in the unstabilized case. We notice that in both cases the cell unknowns can be eliminated locally by using static condensation at each Newton's iteration (4.72).

Remark 4.17 (*Choice of* β_0) To our knowledge, there is no general theory on the choice of β_0 in the case of finite deformations of hyperelastic materials. Following ideas developed in [15, 142, 143], one can consider to take (possibly in an adaptive fashion) the largest eigenvalue (in absolute value) of the elastic modulus $\mathbb{A}$. This choice introduces additional nonlinearities to be handled by Newton's method, and may require some relaxation. Another possibility discussed in [50] for virtual element methods is based on the trace of the Hessian of the isochoric part of the strain-energy density. Such an approach bears similarities with the classic selective integration for FEM, and for Neohookean materials, this choice implies to take $\beta_0 = 1$. Finally, we mention that too large values of the stabilization parameter β_0 can deteriorate the condition number of the stiffness matrix and can cause numerical instabilities in Newton's method. □

4.4 Numerical Examples

We present two examples that are close to industrial simulations: one for linear elasticity, a perforated strip subjected to uniaxial extension, and one for hyperelasticity, the pinching of a pipe. The material parameters are $\mu = 23.3$ MPa and $\lambda = 11650$ MPa, which correspond to a Young modulus $E = 70$ MPa and a Poisson ratio $\nu = 0.499$. Both simulations are in the quasi-incompressible regime to show the robustess of HHO methods.

(a) Euclidean norm of displacement (mm)

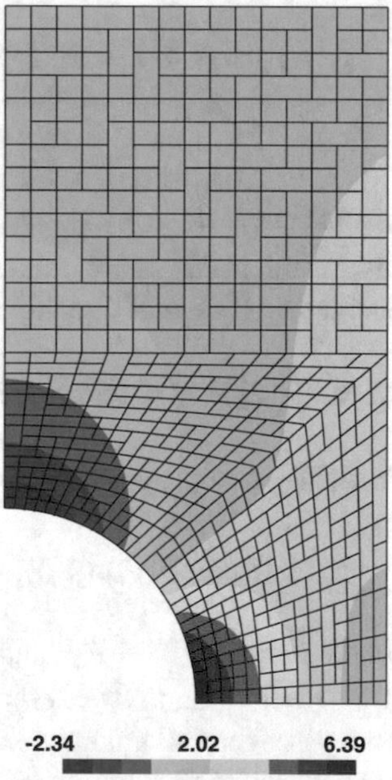

(b) Trace of stress tensor (MPa)

Fig. 4.2 Perforated strip: Euclidean norm of the displacement field (left) and trace of stress tensor (right); polynomial degree $k = 1$; the mesh with hanging nodes is treated as a polygonal mesh and is composed of 536 cells (quadrilaterals, pentagons and hexagons)

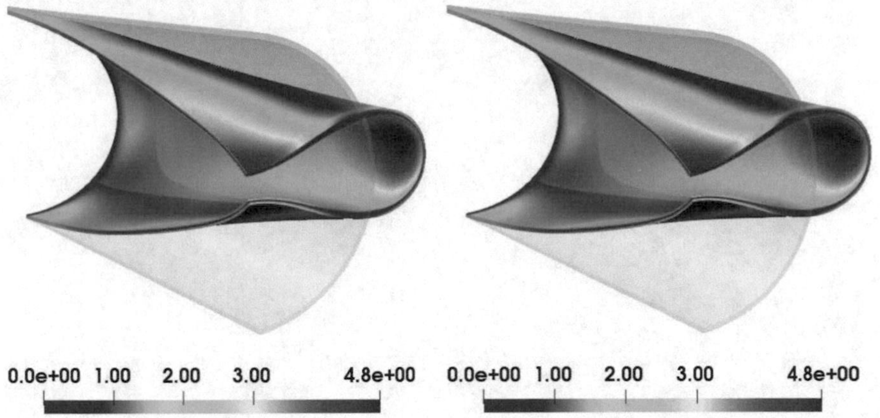

(a) $G^*_{\mathcal{T}} = G_{\mathcal{T}}$ with stabilization (b) $G^*_{\mathcal{T}} = G^{\mathrm{RT}}_{\mathcal{T}}$ without stabilization

Fig. 4.3 Pinching of a pipe: von Mises stress (in MPa) on the deformed configuration for $k = 1$ and different gradient reconstructions

We first consider a strip of width $2L = 200$ mm and height $2H = 360$ mm. The strip is perforated in its middle by a circular hole of radius $R = 50$ mm, and is subjected to a uniaxial extension $\delta = 5$ mm at its top and bottom ends. We consider the linear elasticity model. For symmetry reasons, only one quarter of the strip is discretized. The Euclidean norm of the displacement field and the trace of stress tensor are plotted in Fig. 4.2 for $k = 1$ on a mesh composed of 536 cells with hanging nodes. There is no sign of volumetric locking, thereby comfirming the robustness of HHO methods in the quasi-incompressible limit.

The second numerical example is the pinching of a pipe due to external forces. The pipe has an outer radius of 24 mm and an inner radius of 23 mm (the thickness is equal to 1 mm) and a length equal to 100 mm. One end is clamped, the other end and the inner surface are free, and the outer surface is subjected to a compression force of 0.01 MPa, oriented downwards in the upper half and upwards in the lower half of the outer surface. Since the geometry as well as the boundary conditions are symmetric, it is sufficient to model one half of the pipe in finite deformations. The mesh is composed of 40,500 tetrahedra. The von Mises stress is plotted in Fig. 4.3 for $k = 1$ and different gradient reconstructions (using full polynomials or Raviart–Thomas polynomials with $k = 1$) on the deformed configuration. Both simulations do not present any sign of volumetric locking.

Chapter 5
Elastodynamics

The goal of this chapter is to show how the HHO method can be used for the space semi-discretization of the elastic wave equation. For simplicity, we restrict the scope to media undergoing infinitesimal deformations and governed by a linear stress-strain constitutive relation. We consider first the second-order formulation in time and then the mixed formulation leading to a first-order formulation in time. The time discretization is realized, respectively, by means of Newmark schemes and diagonally-implicit or explicit Runge–Kutta schemes. Interestingly, considering the mixed-order HHO method is instrumental to devise explicit Runge–Kutta schemes. HHO methods for acoustic and elastic wave propagation were developed in [33, 34], and HDG methods for these problems were studied in [63, 127, 133, 140].

5.1 Second-Order Formulation in Time

In this section, we consider the second-order formulation in time of the elastic wave equation. We refer the reader to Sect. 4.1.1 for a description of the linear elasticity model in the case of static problems. Let $J := (0, T_{\mathrm{f}})$ with final time $T_{\mathrm{f}} > 0$ be the time interval, and let Ω be an open, bounded, connected, Lipschitz subset of $\mathbb{R}^d$ in space dimension $d \geq 2$. The elastic wave equation reads as follows:

$$\rho \partial_{tt} \boldsymbol{u} - \nabla \cdot \boldsymbol{\sigma}(\boldsymbol{\varepsilon}(\boldsymbol{u})) = \boldsymbol{f} \quad \text{in } J \times \Omega, \tag{5.1}$$

where ρ is the material density, $\boldsymbol{f}$ the body force, and $\boldsymbol{u}$ the displacement field. The stress tensor $\boldsymbol{\sigma}(\boldsymbol{\varepsilon}(\boldsymbol{u}))$ depends on the displacement field by means of the linearized strain tensor $\boldsymbol{\varepsilon} := \boldsymbol{\varepsilon}(\boldsymbol{u}) := \frac{1}{2}(\nabla \boldsymbol{u} + \nabla \boldsymbol{u}^{\mathsf{T}})$ as follows:

$$\boldsymbol{\sigma}(\boldsymbol{\varepsilon}) := \mathbb{A}\boldsymbol{\varepsilon} := 2\mu\boldsymbol{\varepsilon} + \lambda \operatorname{tr}(\boldsymbol{\varepsilon}) \boldsymbol{I}_d, \tag{5.2}$$

M. Cicuttin et al., *Hybrid High-Order Methods*,
SpringerBriefs in Mathematics,
https://doi.org/10.1007/978-3-030-81477-9_5

where $\mathbb{A}$ is the fourth-order stiffness tensor, μ and λ are the Lamé parameters, and $\boldsymbol{I}_d$ the identity tensor. We assume that the coefficients ρ, μ, and λ are piecewise constant on a partition of Ω into a finite collection of polyhedral subdomains, that ρ and μ take positive values, and that λ takes nonnegative values. The wave equation (5.1) describes the propagation of different types of elastic waves in the medium. In particular, the speeds of P- and S-waves are $c_{\mathrm{P}} := \sqrt{\frac{\lambda+2\mu}{\rho}}$ and $c_{\mathrm{S}} := \sqrt{\frac{\mu}{\rho}}$.

The wave equation (5.1) is subjected to the initial and boundary conditions

$$\boldsymbol{u}(0) = \boldsymbol{u}_0, \ \partial_t \boldsymbol{u}(0) = \boldsymbol{v}_0 \text{ in } \Omega, \qquad \boldsymbol{u} = \boldsymbol{0} \text{ on } J \times \partial\Omega, \tag{5.3}$$

where the homogeneous Dirichlet boundary condition is chosen for simplicity. Assuming that $\boldsymbol{f} \in C^0(\overline{J}; \boldsymbol{L}^2(\Omega))$ with $\overline{J} := [0, T_{\mathrm{f}}]$, $\boldsymbol{u}_0 \in \boldsymbol{H}_0^1(\Omega)$, and $\boldsymbol{v}_0 \in \boldsymbol{H}_0^1(\Omega)$, and that $\boldsymbol{u} \in C^0(\overline{J}; \boldsymbol{H}_0^1(\Omega)) \cap H^2(J; \boldsymbol{L}^2(\Omega))$, we have for a.e. $t \in J$,

$$(\partial_{tt}\boldsymbol{u}(t), \boldsymbol{w})_{\boldsymbol{L}^2(\rho;\Omega)} + a(\boldsymbol{u}(t), \boldsymbol{w}) = (\boldsymbol{f}(t), \boldsymbol{w})_{\boldsymbol{L}^2(\Omega)}, \quad \forall \boldsymbol{w} \in \boldsymbol{H}_0^1(\Omega), \tag{5.4}$$

with the bilinear form

$$a(\boldsymbol{v}, \boldsymbol{w}) := (\boldsymbol{\sigma}(\boldsymbol{\varepsilon}(\boldsymbol{v})), \boldsymbol{\varepsilon}(\boldsymbol{w}))_{\mathbb{L}^2(\Omega)} = (\boldsymbol{\varepsilon}(\boldsymbol{v}), \boldsymbol{\varepsilon}(\boldsymbol{w}))_{\mathbb{L}^2(2\mu;\Omega)} + (\nabla{\cdot}\boldsymbol{v}, \nabla{\cdot}\boldsymbol{w})_{L^2(\lambda;\Omega)}. \tag{5.5}$$

Here, for a weight function $\phi \in L^\infty(\Omega)$ taking nonnegative values, we used the notation $\|v\|_{L^2(\phi;\Omega)} := \|\phi^{\frac{1}{2}} v\|_{L^2(\Omega)}$ for all $v \in L^2(\Omega)$, and a similar notation for vector-valued fields in $\boldsymbol{L}^2(\Omega)$ (this defines a norm if ϕ is uniformly bounded from below away from zero).

An important property of the elastic wave equation is energy balance. The time-dependent energy associated with the weak solution is defined for all $t \in \overline{J}$ as

$$\mathfrak{E}(t) := \frac{1}{2}\|\partial_t \boldsymbol{u}(t)\|^2_{\boldsymbol{L}^2(\rho;\Omega)} + \frac{1}{2}\|\boldsymbol{\varepsilon}(\boldsymbol{u}(t))\|^2_{\mathbb{L}^2(2\mu;\Omega)} + \frac{1}{2}\|\nabla{\cdot}\boldsymbol{u}(t)\|^2_{L^2(\lambda;\Omega)}. \tag{5.6}$$

Assuming $\boldsymbol{u} \in C^1(\overline{J}; \boldsymbol{H}_0^1(\Omega))$ and testing (5.4) against $\boldsymbol{w} := \partial_t \boldsymbol{u}(t)$ gives $\frac{d}{dt}\mathfrak{E}(t) = (\boldsymbol{f}(t), \partial_t \boldsymbol{u}(t))_{\boldsymbol{L}^2(\Omega)}$ for all $t \in J$. Integrating in time over $(0, t)$ leads to the energy balance equation

$$\mathfrak{E}(t) = \mathfrak{E}(0) + \int_0^t (\boldsymbol{f}(s), \partial_t \boldsymbol{u}(s))_{\boldsymbol{L}^2(\Omega)}\, \mathrm{d}s, \tag{5.7}$$

where $\mathfrak{E}(0)$ can be evaluated from the initial condition (5.3). In the absence of body forces, (5.7) implies energy conservation, i.e., $\mathfrak{E}(t) = \mathfrak{E}(0)$ for all $t \in \overline{J}$.

5.1.1 HHO Space Semi-discretization

We consider the discrete setting described in Sect. 4.2.1. In particular, $\mathcal{T}$ is a mesh of Ω belonging to a shape-regular mesh sequence, and we assume that Ω is a polyhedron so that the mesh covers Ω exactly. Moreover, we assume that the mesh is compatible with the above partition of Ω regarding the material properties, so that the parameters ρ, μ, and λ are piecewise constant on the mesh.

To allow for a bit more generality (this will be handy when studying the first-order formulation in time in Sect. 5.2), we consider either the equal-order case or the mixed-order case for the cell and the face unknowns in the HHO method. Letting $k \geq 1$ denote the degree of the face unknowns, the cell unknowns can have degree $k' := k$ (equal-order) or $k' := k + 1$ (mixed-order). Only the equal-order case was considered in Sect. 4.2.1 for the static problem, and we refer the reader to Sect. 3.2.1 for a study of the mixed-order HHO method applied to the Poisson model problem.

We use a unified notation to cover both cases, and for simplicity we use only the superscript k in the HHO spaces composed of polynomial pairs. For every mesh cell $T \in \mathcal{T}$, we set

$$\hat{\boldsymbol{V}}_T^k := \mathbb{P}_d^{k'}(T) \times \mathbb{P}_{d-1}^k(\mathcal{F}_T), \qquad \mathbb{P}_{d-1}^k(\mathcal{F}_T) := \bigtimes_{F \in \mathcal{F}_T} \mathbb{P}_{d-1}^k(F), \tag{5.8}$$

with $\mathbb{P}_d^{k'}(T) := \mathbb{P}_d^{k'}(T; \mathbb{R}^d)$ and $\mathbb{P}_{d-1}^k(F) := \mathbb{P}_{d-1}^k(F; \mathbb{R}^d)$; see Fig. 5.1. A generic element in $\hat{\boldsymbol{V}}_T^k$ is denoted by $\hat{\boldsymbol{v}}_T := (\boldsymbol{v}_T, \boldsymbol{v}_{\partial T})$. The HHO space is then defined as

$$\hat{\boldsymbol{V}}_h^k := \boldsymbol{V}_{\mathcal{T}}^{k'} \times \boldsymbol{V}_{\mathcal{F}}^k, \qquad \boldsymbol{V}_{\mathcal{T}}^{k'} := \bigtimes_{T \in \mathcal{T}} \mathbb{P}_d^{k'}(T), \qquad \boldsymbol{V}_{\mathcal{F}}^k := \bigtimes_{F \in \mathcal{F}} \mathbb{P}_{d-1}^k(F). \tag{5.9}$$

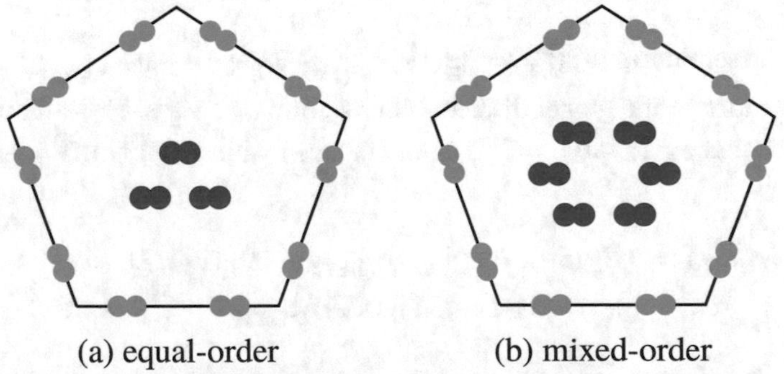

Fig. 5.1 Face (violet) and cell (blue) unknowns in $\hat{\boldsymbol{V}}_T^k$ in a pentagonal cell ($d = 2$) for $k = 1$; left: equal-order case, right: mixed-order case (each dot in a pair represents one basis function associated with one Cartesian component of the displacement)

A generic element in $\hat{V}_h^k$ is denoted by $\hat{v}_h := (v_{\mathcal{T}}, v_{\mathcal{F}})$ with $v_{\mathcal{T}} := (v_T)_{T\in\mathcal{T}}$ and $v_{\mathcal{F}} := (v_F)_{F\in\mathcal{F}}$, and we localize the components of $\hat{v}_h$ associated with a mesh cell $T \in \mathcal{T}$ and its faces by using the notation $\hat{v}_T := \big(v_T, v_{\partial T} := (v_F)_{F\in\mathcal{F}_T}\big) \in \hat{V}_T^k$. The local HHO reduction operator $\hat{I}_T^k : H^1(T) \to \hat{V}_T^k$ is such that $\hat{I}_T^k(v) := (\Pi_T^{k'}(v), \Pi_{\partial T}^k(v_{|\partial T}))$, and its global counterpart $\hat{I}_h^k : H^1(\Omega) \to \hat{V}_h^k$ is such that $\hat{I}_h^k(v) := (\Pi_{\mathcal{T}}^{k'}(v), \Pi_{\mathcal{F}}^k(v_{|\mathcal{F}}))$.

To reconstruct a strain tensor in every mesh cell $T \in \mathcal{T}$, we can consider the displacement reconstruction operator $U_T : \hat{V}_T^k \to \mathbb{P}_d^{k+1}(T)$ defined in (4.26)–(4.27). However, to highlight another possibility which is relevant in the case of nonlinear materials (see, e.g., Sect. 4.3), we consider here a (symmetric-valued) strain reconstruction operator $E_T : \hat{V}_T^k \to \mathbb{P}_d^k(T; \mathbb{R}^{d\times d}_{\mathrm{sym}})$ such that for all $\hat{v}_T \in \hat{V}_T^k$,

$$(E_T(\hat{v}_T), q)_{L^2(T)} = -(v_T, \nabla\cdot q)_{L^2(T)} + (v_{\partial T}, q n_T)_{L^2(\partial T)}, \tag{5.10}$$

for all $q \in \mathbb{P}_d^k(T; \mathbb{R}^{d\times d}_{\mathrm{sym}})$. Notice that $E_T(\hat{v}_T)$ can be evaluated componentwise by inverting the mass matrix associated with a chosen basis of the scalar-valued polynomial space $\mathbb{P}_d^k(T)$. Recalling (4.30) (equal-order case) and (3.9) (mixed-order case), the local stabilization operator is defined as follows:

$$S_{\partial T}(\hat{v}_T) := \Pi_{\partial T}^k\Big(v_{T|\partial T} - v_{\partial T} + \big((I - \Pi_T^k)U_T(\hat{v}_T)\big)_{|\partial T}\Big), \quad \text{if } k' = k, \tag{5.11}$$

$$S_{\partial T}(\hat{v}_T) := \Pi_{\partial T}^k(v_{T|\partial T}) - v_{\partial T}, \qquad \text{if } k' = k+1, \tag{5.12}$$

so that in the equal-order case, the displacement reconstruction operator U_T needs to be evaluated as well. Adapting the proof of Lemma 4.6, one can show that there are $0 < \alpha \le \omega < +\infty$ such that for all $T \in \mathcal{T}$ and all $\hat{v}_T \in \hat{V}_T^k$,

$$\alpha|\hat{v}_T|^2_{\varepsilon,T} \le \|E_T(\hat{v}_T)\|^2_{L^2(T)} + h_T^{-1}\|S_{\partial T}(\hat{v}_T)\|^2_{L^2(\partial T)} \le \omega|\hat{v}_T|^2_{\varepsilon,T}, \tag{5.13}$$

recalling the seminorm $|\hat{v}_T|^2_{\varepsilon,T} := \|\varepsilon(v_T)\|^2_{L^2(T)} + h_T^{-1}\|v_T - v_{\partial T}\|^2_{L^2(\partial T)}$.

We now turn to the global discrete bilinear form $a_h : \hat{V}_h^k \times \hat{V}_h^k \to \mathbb{R}$ which we define such that $a_h(\hat{v}_h, \hat{w}_h) := \sum_{T\in\mathcal{T}_h} a_T(\hat{v}_T, \hat{w}_T)$ with $a_T : \hat{V}_T^k \times \hat{V}_T^k \to \mathbb{R}$ such that

$$\begin{aligned} a_T(\hat{v}_T, \hat{w}_T) := {} & (E_T(\hat{v}_T), E_T(\hat{w}_T))_{L^2(2\mu;T)} + (D_T(\hat{v}_T), D_T(\hat{w}_T))_{L^2(\lambda;T)} \\ & + \tau_{\partial T}(S_{\partial T}(\hat{v}_T), S_{\partial T}(\hat{w}_T))_{L^2(\partial T)}, \end{aligned} \tag{5.14}$$

with the weight $\tau_{\partial T} := 2\mu_{|T} h_T^{-1}$, and where $D_T(\cdot) := \mathrm{tr}(E_T(\cdot))$ coincides with the divergence reconstruction operator defined in (4.28). We define the global strain reconstruction operator $E_{\mathcal{T}} : \hat{V}_h^k \to W_{\mathcal{T}} := \mathbb{P}_d^k(\mathcal{T}; \mathbb{R}^{d\times d}_{\mathrm{sym}})$ such that

$$E_{\mathcal{T}}(\hat{v}_h)_{|T} := E_T(\hat{v}_T), \quad \forall T \in \mathcal{T}, \ \forall \hat{v}_h \in \hat{V}_h^k, \tag{5.15}$$

and the global divergence reconstruction operator $D_{\mathcal{T}} : \hat{\boldsymbol{V}}_h^k \rightarrow \mathbb{P}_d^k(\mathcal{T}; \mathbb{R})$ such that $D_{\mathcal{T}}(\hat{\boldsymbol{v}}_h) := \mathrm{tr}(\boldsymbol{E}_{\mathcal{T}}(\hat{\boldsymbol{v}}_h))$. We also define the global stabilization bilinear form s_h on $\hat{\boldsymbol{V}}_h^k \times \hat{\boldsymbol{V}}_h^k$ such that

$$s_h(\hat{\boldsymbol{v}}_h, \hat{\boldsymbol{w}}_h) := \sum_{T \in \mathcal{T}_h} \tau_{\partial T} (\boldsymbol{S}_{\partial T}(\hat{\boldsymbol{v}}_T), \boldsymbol{S}_{\partial T}(\hat{\boldsymbol{w}}_T))_{L^2(\partial T)}. \tag{5.16}$$

Letting $\boldsymbol{V}_{\mathcal{F},0}^k := \{\boldsymbol{v}_{\mathcal{F}} \in \boldsymbol{V}_{\mathcal{F}}^k \mid \boldsymbol{v}_F = \boldsymbol{0}, \forall F \in \mathcal{F}^{\partial}\}$ and $\hat{\boldsymbol{V}}_{h,0}^k := \boldsymbol{V}_{\mathcal{T}}^k \times \boldsymbol{V}_{\mathcal{F},0}^k$, the space semi-discrete HHO scheme for the elastic wave equation is as follows: Seek $\hat{\boldsymbol{u}}_h := (\boldsymbol{u}_{\mathcal{T}}, \boldsymbol{u}_{\mathcal{F}}) \in C^2(\overline{J}; \hat{\boldsymbol{V}}_{h,0}^k)$ such that for all $t \in \overline{J}$,

$$(\partial_{tt}\boldsymbol{u}_{\mathcal{T}}(t), \boldsymbol{w}_{\mathcal{T}})_{\boldsymbol{L}^2(\rho;\Omega)} + a_h(\hat{\boldsymbol{u}}_h(t), \hat{\boldsymbol{w}}_h) = (\boldsymbol{f}(t), \boldsymbol{w}_{\mathcal{T}})_{\boldsymbol{L}^2(\Omega)}, \tag{5.17}$$

for all $\hat{\boldsymbol{w}}_h := (\boldsymbol{w}_{\mathcal{T}}, \boldsymbol{w}_{\mathcal{F}}) \in \hat{\boldsymbol{V}}_{h,0}^k$. Notice that the acceleration term only involves the cell components; the same remark applies to the body force (as in the static case). Consistently with this observation, the initial conditions for (5.17) only concern $\boldsymbol{u}_{\mathcal{T}}$ and read $\boldsymbol{u}_{\mathcal{T}}(0) = \boldsymbol{\Pi}_{\mathcal{T}}^{k'}(\boldsymbol{u}_0)$, $\partial_t\boldsymbol{u}_{\mathcal{T}}(0) = \boldsymbol{\Pi}_{\mathcal{T}}^{k'}(\boldsymbol{v}_0)$, whereas the boundary condition is encoded in the fact that $\hat{\boldsymbol{u}}_h(t) \in \hat{\boldsymbol{V}}_{h,0}^k$ for all $t \in \overline{J}$. Notice that $\boldsymbol{u}_{\mathcal{F}}(0) \in \boldsymbol{V}_{\mathcal{F},0}^k$ is uniquely determined by the equations $a_h((\boldsymbol{u}_{\mathcal{T}}(0), \boldsymbol{u}_{\mathcal{F}}(0)), (\boldsymbol{0}, \boldsymbol{w}_{\mathcal{F}})) = 0$ for all $\boldsymbol{w}_{\mathcal{F}} \in \boldsymbol{V}_{\mathcal{F},0}^k$ with $\boldsymbol{u}_{\mathcal{T}}(0)$ specified by the initial condition.

The time-dependent energy associated with the space semi-discrete HHO problem (5.17) is defined for all $t \in \overline{J}$ as (compare with (5.6))

$$\begin{aligned} \mathfrak{E}_h(t) := {} & \frac{1}{2}\|\partial_t\boldsymbol{u}_{\mathcal{T}}(t)\|^2_{\boldsymbol{L}^2(\rho;\Omega)} + \frac{1}{2}\|\boldsymbol{E}_{\mathcal{T}}(\hat{\boldsymbol{u}}_h(t))\|^2_{\mathbb{L}^2(2\mu;\Omega)} + \frac{1}{2}\|D_{\mathcal{T}}(\hat{\boldsymbol{u}}_h(t))\|^2_{L^2(\lambda;\Omega)} \\ & + \frac{1}{2}s_h(\hat{\boldsymbol{u}}_h(t), \hat{\boldsymbol{u}}_h(t)). \end{aligned} \tag{5.18}$$

Then, proceeding as in the continuous case, one shows that

$$\mathfrak{E}_h(t) = \mathfrak{E}_h(0) + \int_0^t (\boldsymbol{f}(s), \partial_t\boldsymbol{u}_{\mathcal{T}}(s))_{\boldsymbol{L}^2(\Omega)}\, \mathrm{d}s, \tag{5.19}$$

so that, in the absence of body forces, (5.19) implies again energy conservation, i.e., $\mathfrak{E}_h(t) = \mathfrak{E}_h(0)$ for all $t \in \overline{J}$.

Let $N_{\mathcal{T}}^{k'} := \dim(\boldsymbol{V}_{\mathcal{T}}^{k'})$ and $N_{\mathcal{F},0}^k := \dim(\boldsymbol{V}_{\mathcal{F},0}^k)$. Let $(\mathsf{U}_{\mathcal{T}}(t), \mathsf{U}_{\mathcal{F}}(t)) \in \mathbb{R}^{N_{\mathcal{T}}^{k'} \times N_{\mathcal{F},0}^k}$ be the component vectors of the space semi-discrete solution $\hat{\boldsymbol{u}}_h(t) \in \hat{\boldsymbol{V}}_{h,0}^k$ once bases $\{\boldsymbol{\varphi}_i\}_{1 \le i \le N_{\mathcal{T}}^{k'}}$ and $\{\boldsymbol{\psi}_j\}_{1 \le j \le N_{\mathcal{F},0}^k}$ for $\boldsymbol{V}_{\mathcal{T}}^{k'}$ and $\boldsymbol{V}_{\mathcal{F},0}^k$, respectively, have been chosen. Let $\mathsf{F}_{\mathcal{T}}(t) \in \mathbb{R}^{N_{\mathcal{T}}^{k'}}$ have components given by $\mathsf{F}_i(t) := (\boldsymbol{f}(t), \boldsymbol{\varphi}_i)_{\boldsymbol{L}^2(\Omega)}$ for all $1 \le i \le N_{\mathcal{T}}^{k'}$ and all $t \in \overline{J}$. The algebraic realization of (5.17) is as follows: For all $t \in \overline{J}$,

$$\begin{bmatrix} \mathsf{M}_{\mathcal{T}\mathcal{T}} & 0 \\ 0 & 0 \end{bmatrix} \begin{bmatrix} \partial_{tt}\mathsf{U}_{\mathcal{T}}(t) \\ \bullet \end{bmatrix} + \begin{bmatrix} \mathsf{A}_{\mathcal{T}\mathcal{T}} & \mathsf{A}_{\mathcal{T}\mathcal{F}} \\ \mathsf{A}_{\mathcal{F}\mathcal{T}} & \mathsf{A}_{\mathcal{F}\mathcal{F}} \end{bmatrix} \begin{bmatrix} \mathsf{U}_{\mathcal{T}}(t) \\ \mathsf{U}_{\mathcal{F}}(t) \end{bmatrix} = \begin{bmatrix} \mathsf{F}_{\mathcal{T}}(t) \\ 0 \end{bmatrix}, \tag{5.20}$$

with the mass matrix $\mathsf{M}_{\mathcal{TT}}$ associated with the inner product in $L^2(\rho;\Omega)$ and the cell basis functions, whereas the symmetric positive-definite stiffness matrix with blocks $\mathsf{A}_{\mathcal{TT}}, \mathsf{A}_{\mathcal{TF}}, \mathsf{A}_{\mathcal{FT}}, \mathsf{A}_{\mathcal{FF}}$ is associated with the bilinear form a_h and the cell and face basis functions. The bullet stands for $\partial_{tt}\mathsf{U}_{\mathcal{F}}(t)$ which is irrelevant owing to the structure of the mass matrix. The matrices $\mathsf{M}_{\mathcal{TT}}$ and $\mathsf{A}_{\mathcal{TT}}$ are block-diagonal, but this is not the case for the matrix $\mathsf{A}_{\mathcal{FF}}$ since the components $\boldsymbol{u}_{\mathcal{F}}(t)$ attached to the faces belonging to the same cell are coupled together through the strain reconstruction operator (and the stabilization operator in the equal-order case).

Remark 5.1 (*Error analysis*) The error analysis for the space semi-discrete problem is performed in [34, Theorems 3.1 and 3.2] for the acoustic wave equation and can be extended to the elastic wave equation. Following the seminal ideas from [11, 89], the key idea is to exploit the approximation properties of the discrete solution operator in the static case (see Sect. 4.2.3 for linear elasticity) and use the stability properties of the wave equation in time. For brevity, we only mention that the energy-error $\|\partial_t \boldsymbol{u} - \partial_t \boldsymbol{u}_{\mathcal{T}}\|_{L^\infty(J;\boldsymbol{L}^2(\rho;\Omega))} + \|\boldsymbol{\varepsilon}(\boldsymbol{u}) - \boldsymbol{E}_{\mathcal{T}}(\hat{\boldsymbol{u}}_h)\|_{L^\infty(J;\mathbb{L}^2(2\mu;\Omega))}$ decays as $O(h^{k+1})$ if $\boldsymbol{u} \in C^1(\overline{J};\boldsymbol{H}^{k+2}(\Omega))$, and assuming full elliptic regularity pickup ($s = 1$), $\|\boldsymbol{\Pi}_{\mathcal{T}}^{k'}(\boldsymbol{u}) - \boldsymbol{u}_{\mathcal{T}}\|_{L^\infty(J;\boldsymbol{L}^2(\mu;\Omega))}$ decays as $O(h^{k+2})$ if additionally $\boldsymbol{f} \in C^1(\overline{J};\boldsymbol{H}^k(\Omega))$. □

5.1.2 Time Discretization

Let $(t^n)_{0\le n\le N}$ be the discrete time nodes with $t^0 := 0$ and $t^N := T_{\mathrm{f}}$. For simplicity, we consider a fixed time step $\Delta t := \frac{T_{\mathrm{f}}}{N}$. A classical time discretization of (5.17) relies on the Newmark scheme with parameters β and γ, which is second-order accurate in time, implicit if $\beta > 0$, unconditionally stable if $\frac{1}{2} \le \gamma \le 2\beta$ (the classical choice is $\gamma = \frac{1}{2}$ and $\beta = \frac{1}{4}$) and conditionally stable if $\frac{1}{2} \le \gamma$ and $2\beta < \gamma$. We detail the implementation for $\beta > 0$. The Newmark scheme considers a displacement, a velocity, and an acceleration at each time node, which are all hybrid unknowns, say $\hat{\boldsymbol{u}}_h^n, \hat{\boldsymbol{v}}_h^n, \hat{\boldsymbol{x}}_h^n \in \hat{\boldsymbol{V}}_{h,0}^k$. The scheme is initialized by setting $\hat{\boldsymbol{u}}_h^0 := \hat{\boldsymbol{I}}_h^k(\boldsymbol{u}_0)$, $\hat{\boldsymbol{v}}_h^0 := \hat{\boldsymbol{I}}_h^k(\boldsymbol{v}_0)$, and the initial acceleration $\hat{\boldsymbol{x}}_h^0 \in \hat{\boldsymbol{V}}_{h,0}^k$ is defined by solving $(\boldsymbol{x}_{\mathcal{T}}^0, \boldsymbol{w}_{\mathcal{T}})_{\boldsymbol{L}^2(\rho;\Omega)} + a_h(\hat{\boldsymbol{u}}_h^0, (\boldsymbol{w}_{\mathcal{T}}, \boldsymbol{0})) = (\boldsymbol{f}(0), \boldsymbol{w}_{\mathcal{T}})_{\boldsymbol{L}^2(\Omega)}$ for all $\boldsymbol{w}_{\mathcal{T}} \in \boldsymbol{V}_{\mathcal{T}}^{k'}$ and $a_h(\hat{\boldsymbol{x}}_h^0, (\boldsymbol{0}, \boldsymbol{w}_{\mathcal{F}})) = 0$ for all $\boldsymbol{w}_{\mathcal{F}} \in \boldsymbol{V}_{\mathcal{F},0}^k$. Then, given $\hat{\boldsymbol{u}}_h^n, \hat{\boldsymbol{v}}_h^n, \hat{\boldsymbol{x}}_h^n$ from the previous time-step or the initial condition, the HHO-Newmark scheme proceeds as follows: For all $n \in \{0, \dots, N-1\}$,

1. Predictor step: provisional displacement $\hat{\boldsymbol{u}}_h^{*n} := \hat{\boldsymbol{u}}_h^n + \Delta t \hat{\boldsymbol{v}}_h^n + \frac{1}{2}\Delta t^2(1-2\beta)\hat{\boldsymbol{x}}_h^n$ and provisional velocity $\hat{\boldsymbol{v}}_h^{*n} := \hat{\boldsymbol{v}}_h^n + \Delta t(1-\gamma)\hat{\boldsymbol{x}}_h^n$.
2. Linear solve to find the acceleration $\hat{\boldsymbol{x}}_h^{n+1} \in \hat{\boldsymbol{V}}_{h,0}^k$ such that for all $\hat{\boldsymbol{w}}_h \in \hat{\boldsymbol{V}}_{h,0}^k$,

$$(\boldsymbol{x}_{\mathcal{T}}^{n+1}, \boldsymbol{w}_{\mathcal{T}})_{\boldsymbol{L}^2(\rho;\Omega)} + \beta\Delta t^2 a_h(\hat{\boldsymbol{x}}_h^{n+1}, \hat{\boldsymbol{w}}_h) = (\boldsymbol{f}(t^{n+1}), \boldsymbol{w}_{\mathcal{T}})_{\boldsymbol{L}^2(\Omega)} - a_h(\hat{\boldsymbol{u}}_h^{*n}, \hat{\boldsymbol{w}}_h). \tag{5.21}$$

3. Corrector step: $\hat{\boldsymbol{u}}_h^{n+1} := \hat{\boldsymbol{u}}_h^{*n} + \beta\Delta t^2\hat{\boldsymbol{x}}_h^{n+1}$, $\hat{\boldsymbol{v}}_h^{n+1} := \hat{\boldsymbol{v}}_h^{*n} + \gamma\Delta t\hat{\boldsymbol{x}}_h^{n+1}$.

The algebraic realization of (5.21) amounts to finding $(\mathsf{X}_{\mathcal{T}}^{n+1}, \mathsf{X}_{\mathcal{F}}^{n+1}) \in \mathbb{R}^{N_{\mathcal{T}}^{k'} \times N_{\mathcal{F},0}^{k}}$ such that

$$\left(\begin{bmatrix} \mathsf{M}_{\mathcal{T}\mathcal{T}} & 0 \\ 0 & 0 \end{bmatrix} + \beta\Delta t^2 \begin{bmatrix} \mathsf{A}_{\mathcal{T}\mathcal{T}} & \mathsf{A}_{\mathcal{T}\mathcal{F}} \\ \mathsf{A}_{\mathcal{F}\mathcal{T}} & \mathsf{A}_{\mathcal{F}\mathcal{F}} \end{bmatrix}\right) \begin{bmatrix} \mathsf{X}_{\mathcal{T}}^{n+1} \\ \mathsf{X}_{\mathcal{F}}^{n+1} \end{bmatrix} = \begin{bmatrix} \mathsf{G}_{\mathcal{T}}^{n+1} \\ \mathsf{G}_{\mathcal{F}}^{n+1} \end{bmatrix}, \tag{5.22}$$

with $\mathsf{G}_{\mathcal{T}}^{n+1} := \mathsf{F}_{\mathcal{T}}^{n+1} - (\mathsf{A}_{\mathcal{T}\mathcal{T}}\mathsf{U}_{\mathcal{T}}^{*n} + \mathsf{A}_{\mathcal{T}\mathcal{F}}\mathsf{U}_{\mathcal{F}}^{*n})$, $\mathsf{G}_{\mathcal{F}}^{n+1} := -(\mathsf{A}_{\mathcal{F}\mathcal{T}}\mathsf{U}_{\mathcal{T}}^{*n} + \mathsf{A}_{\mathcal{F}\mathcal{F}}\mathsf{U}_{\mathcal{F}}^{*n})$, and $(\mathsf{U}_{\mathcal{T}}^{*n}, \mathsf{U}_{\mathcal{F}}^{*n})$ are the components of the predicted displacement $\hat{\boldsymbol{u}}_h^{*n}$. Since the matrix $\mathsf{M}_{\mathcal{T}\mathcal{T}} + \beta\Delta t^2\mathsf{A}_{\mathcal{T}\mathcal{T}}$ is block-diagonal, static condensation can be applied to (5.22), i.e., the cell acceleration unknowns can be eliminated locally, leading to a global transmission problem coupling only the face acceleration unknowns.

An important property of the HHO-Newmark scheme is energy balance. For all $n \in \{0, \dots, N\}$, we define the discrete energy

$$\begin{aligned}\mathfrak{E}^n := \frac{1}{2}\|\boldsymbol{v}_{\mathcal{T}}^n\|_{\boldsymbol{L}^2(\rho;\Omega)}^2 + \frac{1}{2}\|\boldsymbol{E}_{\mathcal{T}}(\hat{\boldsymbol{u}}_h^n)\|_{\mathbb{L}^2(2\mu;\Omega)}^2 + \frac{1}{2}\|D_{\mathcal{T}}(\hat{\boldsymbol{u}}_h^n)\|_{L^2(\lambda;\Omega)}^2 \\ + \frac{1}{2}s_h(\hat{\boldsymbol{u}}_h^n, \hat{\boldsymbol{u}}_h^n) + \delta\Delta t^2\|\boldsymbol{x}_{\mathcal{T}}^n\|_{\boldsymbol{L}^2(\rho;\Omega)}^2,\end{aligned} \tag{5.23}$$

with $\delta := \frac{1}{4}(2\beta - \gamma)$, i.e., $\delta = 0$ for the standard choice $\beta = \frac{1}{4}$, $\gamma = \frac{1}{2}$. Using standard manipulations for Newmark schemes (see [33, Lemma 3.3]), one can show that $\mathfrak{E}^n = \mathfrak{E}^1 + \sum_{m=1}^{n-1}\frac{1}{2}(\boldsymbol{f}(t^{m+1}) + \boldsymbol{f}(t^m), \boldsymbol{u}_{\mathcal{T}}^{m+1} - \boldsymbol{u}_{\mathcal{T}}^m)_{\boldsymbol{L}^2(\Omega)}$, so that $\mathfrak{E}^n$ is conserved in the absence of body forces.

5.2 First-Order Formulation in Time

The first-order formulation of the elastic wave equation is obtained by introducing the velocity field $\boldsymbol{v} := \partial_t\boldsymbol{u}$ and the stress tensor $\boldsymbol{\sigma} := \boldsymbol{\sigma}(\boldsymbol{\varepsilon}(\boldsymbol{u}))$ as independent unknowns. Taking the time derivative of (5.2) and exchanging the order of derivatives leads to

$$\mathbb{A}^{-1}\partial_t\boldsymbol{\sigma} - \boldsymbol{\varepsilon}(\boldsymbol{v}) = \boldsymbol{0}, \qquad \rho\partial_t\boldsymbol{v} - \nabla{\cdot}\boldsymbol{\sigma} = \boldsymbol{f}, \tag{5.24}$$

with $\mathbb{A}^{-1}\boldsymbol{\tau} = \frac{1}{2\mu}(\boldsymbol{\tau} - \frac{\lambda}{2\mu+\lambda d}\mathrm{tr}(\boldsymbol{\tau})\boldsymbol{I}_d)$, together with the initial conditions $\boldsymbol{\sigma}(0) = \mathbb{A}\boldsymbol{\varepsilon}(\boldsymbol{u}_0)$, $\boldsymbol{v}(0) = \boldsymbol{v}_0$ in Ω, and the boundary condition $\boldsymbol{v} = \boldsymbol{0}$ on $J \times \Gamma$. Assuming that $\boldsymbol{v} \in H^1(J; \boldsymbol{L}^2(\Omega)) \cap L^2(J; \boldsymbol{H}_0^1(\Omega))$ and $\boldsymbol{\sigma} \in H^1(J; L^2(\Omega; \mathbb{R}_{\mathrm{sym}}^{d\times d}))$, we obtain

$$\begin{cases} (\partial_t\boldsymbol{\sigma}(t), \boldsymbol{\tau})_{\mathbb{L}^2(\mathbb{A}^{-1};\Omega)} - (\boldsymbol{\varepsilon}(\boldsymbol{v}(t)), \boldsymbol{\tau})_{\mathbb{L}^2(\Omega)} = 0, \\ (\partial_t\boldsymbol{v}(t), \boldsymbol{w})_{\boldsymbol{L}^2(\rho;\Omega)} + (\boldsymbol{\sigma}(t), \boldsymbol{\varepsilon}(\boldsymbol{w}))_{\mathbb{L}^2(\Omega)} = (\boldsymbol{f}(t), \boldsymbol{w})_{\boldsymbol{L}^2(\Omega)}, \end{cases} \tag{5.25}$$

for all $(\boldsymbol{\tau}, \boldsymbol{w}) \in L^2(\Omega; \mathbb{R}_{\mathrm{sym}}^{d\times d}) \times \boldsymbol{H}_0^1(\Omega)$ and a.e. $t \in J$.

5.2.1 HHO Space Semi-discretization

The idea is to approximate $\boldsymbol{\sigma}$ by a cellwise unknown $\boldsymbol{\sigma}_{\mathcal{T}} \in C^1(\overline{J}; \boldsymbol{W}_{\mathcal{T}})$ and $\boldsymbol{v}$ by a hybrid unknown $\hat{\boldsymbol{v}}_h \in C^1(\overline{J}; \hat{\boldsymbol{V}}^k_{h,0})$ (recall that $\boldsymbol{W}_{\mathcal{T}} := \mathbb{P}^k_d(\mathcal{T}; \mathbb{R}^{d\times d}_{\mathrm{sym}})$). The space semi-discrete problem then reads as follows: For all $t \in \overline{J}$,

$$\begin{cases} (\partial_t \boldsymbol{\sigma}_{\mathcal{T}}(t), \boldsymbol{\tau}_{\mathcal{T}})_{\boldsymbol{L}^2(\mathbb{A}^{-1};\Omega)} - (\boldsymbol{E}_{\mathcal{T}}(\hat{\boldsymbol{v}}_h(t)), \boldsymbol{\tau}_{\mathcal{T}})_{\boldsymbol{L}^2(\Omega)} = 0, \\ (\partial_t \boldsymbol{v}_{\mathcal{T}}(t), \boldsymbol{w}_{\mathcal{T}})_{\boldsymbol{L}^2(\rho;\Omega)} + (\boldsymbol{\sigma}_{\mathcal{T}}(t), \boldsymbol{E}_{\mathcal{T}}(\hat{\boldsymbol{w}}_h))_{\boldsymbol{L}^2(\Omega)} + \tilde{s}_h(\hat{\boldsymbol{v}}_h(t), \hat{\boldsymbol{w}}_h) = (\boldsymbol{f}(t), \boldsymbol{w}_{\mathcal{T}})_{\boldsymbol{L}^2(\Omega)}, \end{cases} \tag{5.26}$$

for all $(\boldsymbol{\tau}_{\mathcal{T}}, \hat{\boldsymbol{w}}_h) \in \boldsymbol{W}_{\mathcal{T}} \times \hat{\boldsymbol{V}}^k_{h,0}$, where the global strain reconstruction operator $\boldsymbol{E}_{\mathcal{T}}$ is defined in (5.15). The stabilization bilinear form is $\tilde{s}_h(\hat{\boldsymbol{v}}_h, \hat{\boldsymbol{w}}_h) := \sum_{T\in\mathcal{T}_h} \tilde{\tau}_{\partial T}(\boldsymbol{S}_{\partial T}(\hat{\boldsymbol{v}}_T), \boldsymbol{S}_{\partial T}(\hat{\boldsymbol{w}}_T))_{\boldsymbol{L}^2(\partial T)}$ with parameter $\tilde{\tau}_{\partial T} := \rho c_{\mathrm{S}} \frac{\ell_\Omega}{h_T}$ (that is, $\tilde{\tau}_{\partial T} = O(h_T^{-1})$) or $\tilde{\tau}_{\partial T} := \rho c_{\mathrm{S}}$ (that is, $\tilde{\tau}_{\partial T} = O(1)$) where $c_{\mathrm{S}} = \sqrt{\mu/\rho}$. The initial conditions for (5.26) are $\boldsymbol{\sigma}_{\mathcal{T}}(0) = \mathbb{A}\boldsymbol{E}_{\mathcal{T}}(\hat{\boldsymbol{I}}^k_h(\boldsymbol{u}_0))$ and $\boldsymbol{v}_{\mathcal{T}}(0) = \Pi^{k'}_{\mathcal{T}}(\boldsymbol{v}_0)$, whereas the boundary condition is encoded in the fact that $\hat{\boldsymbol{v}}_h(t) \in \hat{\boldsymbol{V}}^k_{h,0}$ for all $t \in \overline{J}$.

The space semi-discrete schemes (5.17) and (5.26) are not equivalent. Indeed, assume that $\hat{\boldsymbol{u}}_h$ solves (5.17), $(\boldsymbol{\sigma}_{\mathcal{T}}, \hat{\boldsymbol{v}}_h)$ solves (5.26), and set $\hat{\boldsymbol{z}}_h(t) := \hat{\boldsymbol{u}}_h(0) + \int_0^t \hat{\boldsymbol{v}}_h(s)\,\mathrm{d}s$. Then, observing that the first equation in (5.26) implies that $\partial_t \boldsymbol{\sigma}_{\mathcal{T}}(t) = \mathbb{A}\boldsymbol{E}_{\mathcal{T}}(\hat{\boldsymbol{v}}_h(t))$, using the initial condition for $\boldsymbol{\sigma}_{\mathcal{T}}$ and the linearity of $\boldsymbol{E}_{\mathcal{T}}$ gives $\boldsymbol{\sigma}_{\mathcal{T}}(t) = \mathbb{A}\boldsymbol{E}_{\mathcal{T}}(\hat{\boldsymbol{u}}_h(0)) + \int_0^t \mathbb{A}\boldsymbol{E}_{\mathcal{T}}(\hat{\boldsymbol{v}}_h(s))\,\mathrm{d}s = \mathbb{A}\boldsymbol{E}_{\mathcal{T}}(\hat{\boldsymbol{u}}_h(t))$ for all $t \in \overline{J}$. Substituting into the second equation in (5.26) and since $\partial_t \boldsymbol{v}_{\mathcal{T}} = \partial_{tt} \boldsymbol{z}_{\mathcal{T}}$, we infer that for all $\hat{\boldsymbol{w}}_h \in \hat{\boldsymbol{V}}^k_{h,0}$,

$$\begin{aligned} (\partial_{tt} \boldsymbol{z}_{\mathcal{T}}(t), \boldsymbol{w}_{\mathcal{T}})_{\boldsymbol{L}^2(\rho;\Omega)} + (\boldsymbol{E}_{\mathcal{T}}(\hat{\boldsymbol{z}}_h(t)), \boldsymbol{E}_{\mathcal{T}}(\hat{\boldsymbol{w}}_h))_{\boldsymbol{L}^2(\mathbb{A};\Omega)} + \tilde{s}_h(\partial_t \hat{\boldsymbol{z}}_h(t), \hat{\boldsymbol{w}}_h) \\ = (\boldsymbol{f}(t), \boldsymbol{w}_{\mathcal{T}})_{\boldsymbol{L}^2(\Omega)}, \end{aligned}$$

which differs from (5.17) in the form of the stabilization term. This difference in structure between the two formulations has an impact on energy conservation. Indeed, defining the discrete energy for all $t \in \overline{J}$ as

$$\mathfrak{E}^*_h(t) := \frac{1}{2}\|\boldsymbol{v}_{\mathcal{T}}(t)\|^2_{\boldsymbol{L}^2(\rho;\Omega)} + \frac{1}{2}\|\boldsymbol{\sigma}_{\mathcal{T}}(t)\|^2_{\boldsymbol{L}^2(\mathbb{A}^{-1};\Omega)}, \tag{5.27}$$

testing (5.26) with $(\boldsymbol{\tau}_{\mathcal{T}}, \hat{\boldsymbol{w}}_h) := (\boldsymbol{\sigma}_{\mathcal{T}}(t), \hat{\boldsymbol{v}}_h(t))$ for all $t \in J$ and integrating over time leads to

$$\mathfrak{E}^*_h(t) + \int_0^t \tilde{s}_h(\hat{\boldsymbol{v}}_h(s), \hat{\boldsymbol{v}}_h(s))\,\mathrm{d}s = \mathfrak{E}^*_h(0) + \int_0^t (\boldsymbol{f}(s), \boldsymbol{v}_{\mathcal{T}}(s))_{\boldsymbol{L}^2(\Omega)}\,\mathrm{d}s. \tag{5.28}$$

Comparing with (5.19), we see that in the second-order formulation, the stabilization is included in the definition of the discrete energy and an exact energy balance is obtained, whereas in the first-order formulation, the discrete energy is independent of the stabilization, but the latter plays a dissipative role in the energy balance.

Remark 5.2 (*Link with HDG*) Recalling the material in Sect. 1.5.2, the space semi-discrete problem (5.26) can be rewritten as an HDG formulation for the first-order wave equation. HDG methods typically consider the stabilization parameter $\tilde{\tau}_{\partial T} = O(1)$ [127]; see also [140, Table 4] for a numerical study. □

Let $M_{\mathcal{T}}^k := \dim(\boldsymbol{W}_{\mathcal{T}}) = \frac{d(d+1)}{2} N_{\mathcal{T}}^k$ and $\{\boldsymbol{\zeta}_i\}_{1 \le i \le M_{\mathcal{T}}^k}$ be the chosen basis for $\boldsymbol{W}_{\mathcal{T}}$. Let $\mathsf{Z}_{\mathcal{T}}(t) \in \mathbb{R}^{M_{\mathcal{T}}^k}$ and $(\mathsf{V}_{\mathcal{T}}(t), \mathsf{V}_{\mathcal{F}}(t)) \in \mathbb{R}^{N_{\mathcal{T}}^{k'} \times N_{\mathcal{F},0}^k}$ be the component vectors of $\boldsymbol{\sigma}_{\mathcal{T}}(t) \in \boldsymbol{W}_{\mathcal{T}}$ and $\hat{\boldsymbol{v}}_h(t) \in \hat{\boldsymbol{V}}_{h,0}^k$, respectively. Let $\mathsf{M}_{\mathcal{T}\mathcal{T}}^{\sigma}$ be the mass matrix associated with the inner product in $\boldsymbol{L}^2(\mathbb{A}^{-1}; \Omega)$ and the basis functions $\{\boldsymbol{\zeta}_i\}_{1 \le i \le M_{\mathcal{T}}^k}$, and recall that $\mathsf{M}_{\mathcal{T}\mathcal{T}}$ is the mass matrix associated with the inner product in $L^2(\rho; \Omega)$ and the basis functions $\{\boldsymbol{\varphi}_i\}_{1 \le i \le N_{\mathcal{T}}^{k'}}$. Let $\mathsf{S}_{\mathcal{T}\mathcal{T}}, \mathsf{S}_{\mathcal{T}\mathcal{F}}, \mathsf{S}_{\mathcal{F}\mathcal{T}}, \mathsf{S}_{\mathcal{F}\mathcal{F}}$ be the four blocks composing the matrix representing the stabilization bilinear form $\tilde{s}_h$. Let $\mathsf{K}_{\mathcal{T}} \in \mathbb{R}^{M_{\mathcal{T}}^k \times N_{\mathcal{T}}^{k'}}$ and $\mathsf{K}_{\mathcal{F}} \in \mathbb{R}^{M_{\mathcal{T}}^k \times N_{\mathcal{F},0}^k}$ be the (rectangular) matrices representing the strain reconstruction operator $\boldsymbol{E}_{\mathcal{T}}$. The algebraic realization of (5.26) is as follows: For all $t \in \overline{J}$,

$$\begin{bmatrix} \mathsf{M}_{\mathcal{T}\mathcal{T}}^{\sigma} & 0 & 0 \\ 0 & \mathsf{M}_{\mathcal{T}\mathcal{T}} & 0 \\ 0 & 0 & 0 \end{bmatrix} \begin{bmatrix} \partial_t \mathsf{Z}_{\mathcal{T}}(t) \\ \partial_t \mathsf{V}_{\mathcal{T}}(t) \\ \bullet \end{bmatrix} + \begin{bmatrix} 0 & -\mathsf{K}_{\mathcal{T}} & -\mathsf{K}_{\mathcal{F}} \\ \mathsf{K}_{\mathcal{T}}^{\mathsf{T}} & \mathsf{S}_{\mathcal{T}\mathcal{T}} & \mathsf{S}_{\mathcal{T}\mathcal{F}} \\ \mathsf{K}_{\mathcal{F}}^{\mathsf{T}} & \mathsf{S}_{\mathcal{F}\mathcal{T}} & \mathsf{S}_{\mathcal{F}\mathcal{F}} \end{bmatrix} \begin{bmatrix} \mathsf{Z}_{\mathcal{T}}(t) \\ \mathsf{V}_{\mathcal{T}}(t) \\ \mathsf{V}_{\mathcal{F}}(t) \end{bmatrix} = \begin{bmatrix} 0 \\ \mathsf{F}_{\mathcal{T}}(t) \\ 0 \end{bmatrix}, \tag{5.29}$$

where the bullet stands for $\partial_t \mathsf{V}_{\mathcal{F}}(t)$ which is irrelevant owing to the structure of the mass matrix. Notice that the third equation in (5.29) implies that

$$\mathsf{S}_{\mathcal{F}\mathcal{F}} \mathsf{V}_{\mathcal{F}}(t) = -(\mathsf{K}_{\mathcal{F}}^{\mathsf{T}} \mathsf{Z}_{\mathcal{T}}(t) + \mathsf{S}_{\mathcal{F}\mathcal{T}} \mathsf{V}_{\mathcal{T}}(t)), \tag{5.30}$$

and that the submatrix $\mathsf{S}_{\mathcal{F}\mathcal{F}}$ is symmetric positive-definite. A crucial observation is that this submatrix is additionally block-diagonal in the mixed-order case, but this property is lost in the equal-order case owing to the presence of the displacement reconstruction operator in the stabilization (see (5.11)).

Remark 5.3 (*Error analysis*) The error analysis for the space semi-discrete problem (5.26) is performed in [34, Theorem 4.3] for $\tilde{\tau}_{\partial T} = O(h_T^{-1})$ and the acoustic wave equation (it can be extended to the elastic wave equation). In particular, the energy-error $\|\boldsymbol{v} - \boldsymbol{v}_{\mathcal{T}}\|_{L^\infty(J; \boldsymbol{L}^2(\rho;\Omega))} + \|\boldsymbol{\sigma} - \boldsymbol{\sigma}_{\mathcal{T}}\|_{L^\infty(J; \mathbb{L}^2(\mathbb{A}^{-1};\Omega))}$ decays as $O(h^{k+1})$ if $(\boldsymbol{\sigma}, \boldsymbol{v}) \in C^1(\overline{J}; \mathbb{H}^{k+1}(\Omega) \times \boldsymbol{H}^{k+2}(\Omega))$. We refer the reader to [67] for the error analysis in the HDG setting with $\tilde{\tau}_{\partial T} = O(1)$, including an improved $L^\infty(J; \boldsymbol{L}^2(\Omega))$-estimate on a post-processed displacement field decaying at rate $O(h^{k+2})$. □

5.2.2 *Time Discretization*

The space semi-discrete problem (5.26) can be discretized in time by means of a Runge–Kutta (RK) scheme. RK schemes are defined by a set of coefficients, $\{a_{ij}\}_{1 \le i,j \le s}$, $\{b_i\}_{1 \le i \le s}$, $\{c_i\}_{1 \le i \le s}$, where $s \ge 1$ is the number of stages. We consider diagonally implicit RK schemes (DIRK), where $a_{ij} = 0$ if $j > i$, and explicit RK schemes (ERK), where additionally $a_{ii} = 0$. The implementation of DIRK and ERK schemes is slightly different owing to the treatment of the face unknowns.

Let us start with DIRK schemes. For all $n \in \{1, \dots, N\}$, given $(\mathsf{Z}_{\mathcal{T}}^{n-1}, \mathsf{V}_{\mathcal{T}}^{n-1})$ from the previous time-step or the initial condition and letting $\mathsf{F}_{\mathcal{T}}^{n-1+c_j} := \mathsf{F}_{\mathcal{T}}(t_{n-1} + c_j \Delta t)$ for all $1 \le j \le s$, one proceeds as follows:

1. Solve sequentially for all $1 \le i \le s$,

$$\begin{bmatrix} \mathsf{M}_{\mathcal{T}\mathcal{T}}^{\sigma} & 0 & 0 \\ 0 & \mathsf{M}_{\mathcal{T}\mathcal{T}} & 0 \\ 0 & 0 & 0 \end{bmatrix} \begin{bmatrix} \mathsf{Z}_{\mathcal{T}}^{n,i} \\ \mathsf{V}_{\mathcal{T}}^{n,i} \\ \bullet \end{bmatrix} = \begin{bmatrix} \mathsf{M}_{\mathcal{T}\mathcal{T}}^{\sigma} & 0 & 0 \\ 0 & \mathsf{M}_{\mathcal{T}\mathcal{T}} & 0 \\ 0 & 0 & 0 \end{bmatrix} \begin{bmatrix} \mathsf{Z}_{\mathcal{T}}^{n-1} \\ \mathsf{V}_{\mathcal{T}}^{n-1} \\ \bullet \end{bmatrix} + \Delta t \sum_{j=1}^{i} a_{ij} \left(\begin{bmatrix} 0 \\ \mathsf{F}_{\mathcal{T}}^{n-1+c_j} \\ 0 \end{bmatrix} - \begin{bmatrix} 0 & -\mathsf{K}_{\mathcal{T}} & -\mathsf{K}_{\mathcal{F}} \\ \mathsf{K}_{\mathcal{T}}^{\mathsf{T}} & \mathsf{S}_{\mathcal{T}\mathcal{T}} & \mathsf{S}_{\mathcal{T}\mathcal{F}} \\ \mathsf{K}_{\mathcal{F}}^{\mathsf{T}} & \mathsf{S}_{\mathcal{F}\mathcal{T}} & \mathsf{S}_{\mathcal{F}\mathcal{F}} \end{bmatrix} \begin{bmatrix} \mathsf{Z}_{\mathcal{T}}^{n,j} \\ \mathsf{V}_{\mathcal{T}}^{n,j} \\ \mathsf{V}_{\mathcal{F}}^{n,j} \end{bmatrix} \right). \quad (5.31)$$

   This is a linear system for the triple $(\mathsf{Z}_{\mathcal{T}}^{n,i}, \mathsf{V}_{\mathcal{T}}^{n,i}, \mathsf{V}_{\mathcal{F}}^{n,i})$ (which appears on both the left- and right-hand sides). The upper 2×2 submatrix associated with the cell unknowns $(\mathsf{Z}_{\mathcal{T}}^{n,i}, \mathsf{V}_{\mathcal{T}}^{n,i})$ being block-diagonal, static condensation can be efficiently performed in (5.31) leading to a global transmission problem coupling only the components of $\mathsf{V}_{\mathcal{F}}^{n,i}$.
2. Finally set

$$\begin{bmatrix} \mathsf{M}_{\mathcal{T}\mathcal{T}}^{\sigma} & 0 \\ 0 & \mathsf{M}_{\mathcal{T}\mathcal{T}} \end{bmatrix} \begin{bmatrix} \mathsf{Z}_{\mathcal{T}}^{n} \\ \mathsf{V}_{\mathcal{T}}^{n} \end{bmatrix} := \begin{bmatrix} \mathsf{M}_{\mathcal{T}\mathcal{T}}^{\sigma} & 0 \\ 0 & \mathsf{M}_{\mathcal{T}\mathcal{T}} \end{bmatrix} \begin{bmatrix} \mathsf{Z}_{\mathcal{T}}^{n-1} \\ \mathsf{V}_{\mathcal{T}}^{n-1} \end{bmatrix} + \Delta t \sum_{j=1}^{s} b_j \left(\begin{bmatrix} 0 \\ \mathsf{F}_{\mathcal{T}}^{n-1+c_j} \end{bmatrix} - \begin{bmatrix} 0 & -\mathsf{K}_{\mathcal{T}} & -\mathsf{K}_{\mathcal{F}} \\ \mathsf{K}_{\mathcal{T}}^{\mathsf{T}} & \mathsf{S}_{\mathcal{T}\mathcal{T}} & \mathsf{S}_{\mathcal{T}\mathcal{F}} \end{bmatrix} \begin{bmatrix} \mathsf{Z}_{\mathcal{T}}^{n,j} \\ \mathsf{V}_{\mathcal{T}}^{n,j} \\ \mathsf{V}_{\mathcal{F}}^{n,j} \end{bmatrix} \right). \quad (5.32)$$

For ERK schemes, instead, one proceeds as follows:

1. Set $(\mathsf{Z}_{\mathcal{T}}^{n,1}, \mathsf{V}_{\mathcal{T}}^{n,1}) := (\mathsf{Z}_{\mathcal{T}}^{n-1}, \mathsf{V}_{\mathcal{T}}^{n-1})$ and solve $\mathsf{S}_{\mathcal{F}\mathcal{F}} \mathsf{V}_{\mathcal{F}}^{n,1} = -(\mathsf{K}_{\mathcal{F}}^{\mathsf{T}} \mathsf{Z}_{\mathcal{T}}^{n,1} + \mathsf{S}_{\mathcal{F}\mathcal{T}} \mathsf{V}_{\mathcal{T}}^{n,1})$.
2. If $s \ge 2$, solve sequentially for all $2 \le i \le s$,

$$\begin{bmatrix} \mathsf{M}_{\mathcal{T}\mathcal{T}}^{\sigma} & 0 \\ 0 & \mathsf{M}_{\mathcal{T}\mathcal{T}} \end{bmatrix} \begin{bmatrix} \mathsf{Z}_{\mathcal{T}}^{n,i} \\ \mathsf{V}_{\mathcal{T}}^{n,i} \end{bmatrix} = \begin{bmatrix} \mathsf{M}_{\mathcal{T}\mathcal{T}}^{\sigma} & 0 \\ 0 & \mathsf{M}_{\mathcal{T}\mathcal{T}} \end{bmatrix} \begin{bmatrix} \mathsf{Z}_{\mathcal{T}}^{n-1} \\ \mathsf{V}_{\mathcal{T}}^{n-1} \end{bmatrix} + \Delta t \sum_{j=1}^{i-1} a_{ij} \left(\begin{bmatrix} 0 \\ \mathsf{F}_{\mathcal{T}}^{n-1+c_j} \end{bmatrix} - \begin{bmatrix} 0 & -\mathsf{K}_{\mathcal{T}} & -\mathsf{K}_{\mathcal{F}} \\ \mathsf{K}_{\mathcal{T}}^{\mathsf{T}} & \mathsf{S}_{\mathcal{T}\mathcal{T}} & \mathsf{S}_{\mathcal{T}\mathcal{F}} \end{bmatrix} \begin{bmatrix} \mathsf{Z}_{\mathcal{T}}^{n,j} \\ \mathsf{V}_{\mathcal{T}}^{n,j} \\ \mathsf{V}_{\mathcal{F}}^{n,j} \end{bmatrix} \right), \quad (5.33)$$

   and $\mathsf{S}_{\mathcal{F}\mathcal{F}} \mathsf{V}_{\mathcal{F}}^{n,i} = -(\mathsf{K}_{\mathcal{F}}^{\mathsf{T}} \mathsf{Z}_{\mathcal{T}}^{n,i} + \mathsf{S}_{\mathcal{F}\mathcal{T}} \mathsf{V}_{\mathcal{T}}^{n,i})$.
3. Finally update the cell unknowns as in (5.32).

We emphasize that the ERK scheme is effective only in the mixed-order case since the submatrix $\mathsf{S}_{\mathcal{F}\mathcal{F}}$ is then block-diagonal. The HHO-ERK scheme is subjected for its stability to a CFL condition on the time-step. The choice $\tilde{\tau}_{\partial T} = O(1)$ is recommended for the stabilization parameter since it leads to a CFL condition scaling linearly with the mesh size (the scaling is quadratic for $\tilde{\tau}_{\partial T} = O(h_T^{-1})$). For the HHO-DIRK scheme, both choices for the stabilization parameter are viable, and numerical experiments indicate that the choice $\tilde{\tau}_{\partial T} = O(h_T^{-1})$ leads to more accurate solutions, with an $O(h^{k+2})$ decay rate for the $L^\infty(J; \boldsymbol{L}^2(\Omega))$-norm.

5.3 Numerical Example

To illustrate the HHO methods described in the previous sections, we consider the propagation of an elastic wave in a two-dimensional heterogeneous domain Ω such that $\overline{\Omega} = \overline{\Omega}_1 \cup \overline{\Omega}_2$ with $\Omega_1 := (-\frac{3}{2}, \frac{3}{2}) \times (-\frac{3}{2}, 0)$ and $\Omega_2 := (-\frac{3}{2}, \frac{3}{2}) \times (0, \frac{3}{2})$. The material properties are $\rho_1 := 1, \mu_1 = \lambda_1 := 1$ in Ω_1 and $\rho_2 := 1, \mu_2 = \lambda_2 := 9$ in Ω_2, so that $c_{\mathrm{S},2} := 3c_{\mathrm{S},1}$, $c_{\mathrm{P},2} := 3c_{\mathrm{P},1}$. The simulation time is $T_{\mathrm{f}} := 1$, and homogeneous Dirichlet boundary conditions are enforced. The body force is $\boldsymbol{f} := \boldsymbol{0}$, and the initial conditions are $\boldsymbol{u}_0 := \boldsymbol{0}$ together with

$$\boldsymbol{v}_0(x, y) := \theta \exp\big(- \pi^2 \tfrac{r^2}{\lambda^2}\big)(\boldsymbol{x} - \boldsymbol{x}_c), \tag{5.34}$$

with $\theta := 10^{-2}$ [s^{-1}], $\lambda := \frac{v_{\mathrm{P},2}}{f_c}$ [m] with $f_c := 10$ [s^{-1}], $r^2 := \|\boldsymbol{x} - \boldsymbol{x}_c\|_{\ell^2}^2$, $\boldsymbol{x}_c := (0, \frac{2}{3})$. The initial condition corresponds to a Ricker wave centered at the point $\boldsymbol{x}_c \in \Omega_2$. The wave first propagates in Ω_2, then is partially transmitted to Ω_1 and later it is also reflected at the boundary of Ω.

Numerical results are obtained using the Newmark scheme (with $\beta = \frac{1}{4}, \gamma = \frac{1}{2}$), a three-stage singly diagonally implicit RK of order 4 (in short, SDIRK(3, 4)), and a four-stage explicit RK scheme of order 4 (in short, ERK(4)). The Butcher tableaux for the RK schemes are, respectively,

$$\begin{array}{c|ccc} \gamma & \gamma & 0 & 0 \\ \frac{1}{2} & \frac{1}{2}-\gamma & \gamma & 0 \\ 1-\gamma & 2\gamma & 1-4\gamma & \gamma \\ \hline & \delta & 1-2\delta & \delta \end{array} \qquad \begin{array}{c|cccc} 0 & 0 & 0 & 0 & 0 \\ \frac{1}{2} & \frac{1}{2} & 0 & 0 & 0 \\ \frac{1}{2} & 0 & \frac{1}{2} & 0 & 0 \\ 1 & 0 & 0 & 1 & 0 \\ \hline & \frac{1}{6} & \frac{1}{3} & \frac{1}{3} & \frac{1}{6} \end{array} \tag{5.35}$$

with $\gamma := \frac{1}{\sqrt{3}} \cos\big(\frac{\pi}{18}\big) + \frac{1}{2}$, $\delta := \frac{1}{6(2\gamma-1)^2}$. We consider a quadrangular mesh of size $h := 2^{-6}$ and a time-step $\Delta t := 0.1 \times 2^{-6}$. Figure 5.2 reports the velocity profiles over the computational domain at the four simulation times $t \in \{\frac{1}{8}, \frac{1}{4}, \frac{1}{2}, 1\}$. These profiles are obtained using the SDIRK(3, 4) scheme ($k' = k$, $\tilde{\tau}_{\partial T} = O(1)$). We

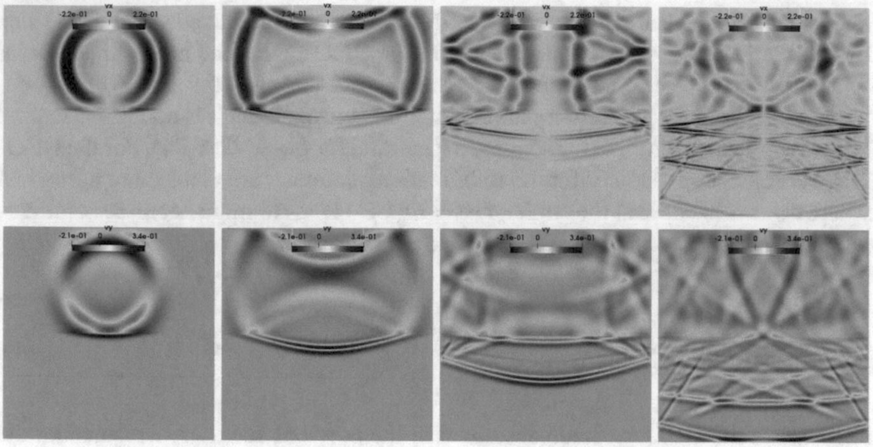

Fig. 5.2 Velocity profiles at the times $t \in \{\frac{1}{8}, \frac{1}{4}, \frac{1}{2}, 1\}$ (from left to right). Upper row: v_x; bottom row: v_y. SDIRK(3, 4) scheme, $k' = k$, $\tilde{\tau}_{\partial T} = O(1)$

observe the various reflections of the elastic waves at the interface and at the domain boundary.

These results can be compared against semi-analytical solutions obtained using the `gar6more2d` software.[1] The semi-analytical solution is based on a reformulation of the problem with zero initial conditions and a Dirac source term with a time delay of 0.15 [s] (this value is tuned to match the choice of the parameter θ, see [18]). The comparisons are made by tracking the velocity at two sensors, one located in Ω_1 at the point $S_1 := (\frac{1}{3}, -\frac{1}{3})$ and one located in Ω_2 at the point $S_2 := (\frac{1}{3}, \frac{1}{3})$. Since the semi-analytical solution assumes propagation in two half-spaces, the comparison with the simulations remains meaningful until the reflected waves at the boundary reach one of the sensors (this happens around the times 0.7 for S_1 and 0.45 for S_2). Figure 5.3 reports the results for the cell velocity component v_x with $\Delta t := 0.1 \times 2^{-8}$ for the second-order Newmark scheme, $\Delta t := 0.1 \times 2^{-6}$ for the SDIRK(3, 4) scheme, and $\Delta t := 0.1 \times 2^{-9}$ for the ERK(4) scheme (owing to the stability condition). Equal-order is used for the Newmark and SDIRK schemes, and mixed-order for the ERK scheme. For both RK schemes, the stabilization parameter is $\tilde{\tau}_{\partial T} = O(1)$. We observe that increasing the polynomial degree in the HHO discretization is beneficial for all the time-stepping schemes, and that the predictions overlap with the semi-analytical solution for $k = 3$. For the RK schemes, the predictions are already quite accurate for $k = 2$, but this is not the case for the Newmark scheme. As expected, the profiles at the sensor S_1 are more difficult to capture due to the transmission of the incoming wave across the interface separating the two media.

[1] https://gforge.inria.fr/projects/gar6more2d/.

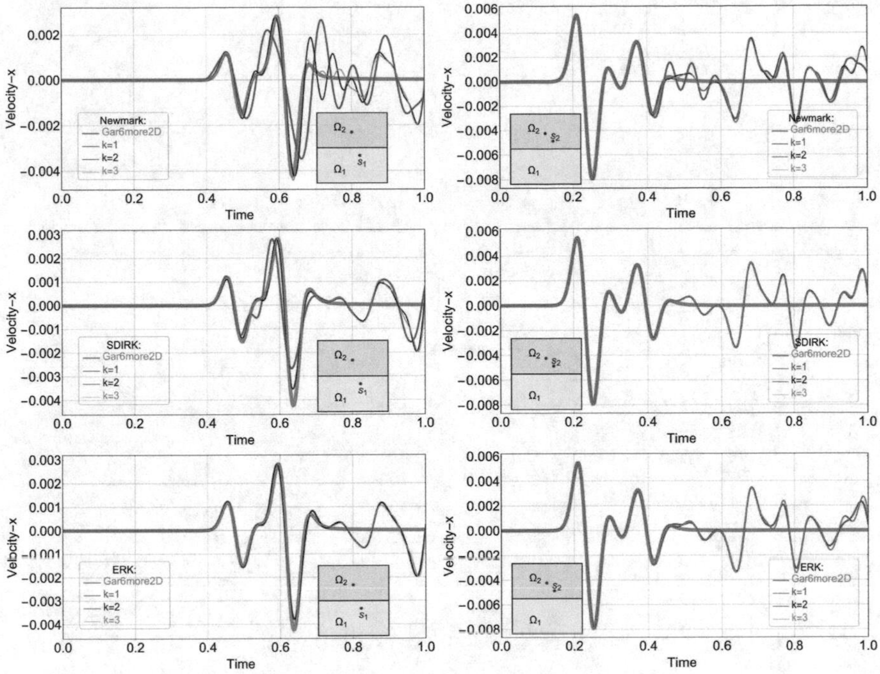

Fig. 5.3 Velocity component v_x at the sensor S_1 (left column) and at the sensor S_2 (right column). The polynomial degree is $k \in \{1, 2, 3\}$. Upper row: Newmark ($\Delta t := 0.1 \times 2^{-8}$), middle row: SDIRK(3, 4) ($\Delta t := 0.1 \times 2^{-6}$), bottom row: ERK(4) ($\Delta t := 0.1 \times 2^{-9}$)

Chapter 6
Contact and Friction

In this chapter, we show how the HHO method can be used to discretize a linear elasticity problem with nonlinear boundary conditions resulting from contact and friction. The main idea is to use a boundary penalty technique to enforce these conditions. This approach leads, under some assumptions, to a discrete semilinear form enjoying a monotonicity property. The error analysis reveals that the degree of the face unknowns on the contact/friction boundary has to be raised to $(k+1)$ to ensure optimal estimates.

6.1 Model Problem

As in Sect. 4.1.1, we consider an elastic body occupying the bounded Lipschitz domain $\Omega \subset \mathbb{R}^d$, $d \in \{2, 3\}$, in the reference configuration. The boundary $\partial\Omega$ is now partitioned into three disjoint subsets: the Dirichlet boundary $\partial\Omega_\mathrm{D}$, the Neumann boundary $\partial\Omega_\mathrm{N}$, and the contact/friction boundary $\partial\Omega_\mathrm{C}$, with $\mathrm{meas}(\partial\Omega_\mathrm{D}) > 0$ (to prevent rigid-body motions) and $\mathrm{meas}(\partial\Omega_\mathrm{C}) > 0$. The body undergoes infinitesimal deformations due to volume forces $\boldsymbol{f} \in L^2(\Omega; \mathbb{R}^d)$ and surface loads $\boldsymbol{g}_\mathrm{N} \in L^2(\partial\Omega_\mathrm{N}; \mathbb{R}^d)$, and it is clamped on $\partial\Omega_\mathrm{D}$ (for simplicity). Recall that the linearized strain tensor associated with a displacement field $\boldsymbol{v} : \Omega \to \mathbb{R}^d$ is $\boldsymbol{\varepsilon}(\boldsymbol{v}) := \frac{1}{2}(\nabla \boldsymbol{v} + \nabla \boldsymbol{v}^\mathsf{T}) \in \mathbb{R}^{d\times d}_\mathrm{sym}$. Assuming a linear elastic behaviour, the Cauchy stress tensor resulting from a strain tensor $\boldsymbol{\varepsilon}$ is given by

$$\boldsymbol{\sigma}(\boldsymbol{\varepsilon}) = 2\mu\boldsymbol{\varepsilon} + \lambda\,\mathrm{tr}(\boldsymbol{\varepsilon})\boldsymbol{I}_d \in \mathbb{R}^{d\times d}_\mathrm{sym}, \tag{6.1}$$

where μ and λ are the Lamé coefficients of the material satisfying $\mu > 0$ and $3\lambda + 2\mu > 0$, and $\boldsymbol{I}_d$ is the identity tensor of order d.

M. Cicuttin et al., *Hybrid High-Order Methods*,
SpringerBriefs in Mathematics,
https://doi.org/10.1007/978-3-030-81477-9_6

Let $\boldsymbol{n}$ be the unit outward normal vector to Ω. On the boundary, we consider the following decompositions into normal and tangential components of a displacement field $\boldsymbol{v}$ and a stress tensor $\boldsymbol{\sigma}$:

$$\boldsymbol{v} = v_n \boldsymbol{n} + \boldsymbol{v}_t \quad \text{and} \quad \boldsymbol{\sigma}_n := \boldsymbol{\sigma n} = \sigma_n \boldsymbol{n} + \boldsymbol{\sigma}_t, \tag{6.2}$$

where $v_n := \boldsymbol{v}{\cdot}\boldsymbol{n}$ and $\sigma_n := \boldsymbol{\sigma}_n{\cdot}\boldsymbol{n}$ (so that $\boldsymbol{v}_t{\cdot}\boldsymbol{n} = 0$ and $\boldsymbol{\sigma}_t{\cdot}\boldsymbol{n} = 0$). The model problem consists in finding the displacement field $\boldsymbol{u} : \Omega \to \mathbb{R}^d$ such that, using the shorthand notation $\boldsymbol{\sigma}(\boldsymbol{u}) := \boldsymbol{\sigma}(\boldsymbol{\varepsilon}(\boldsymbol{u}))$,

$$-\nabla{\cdot}\boldsymbol{\sigma}(\boldsymbol{u}) = \boldsymbol{f} \text{ in } \Omega, \tag{6.3}$$

$$\boldsymbol{u} = \boldsymbol{0} \text{ on } \partial\Omega_{\mathrm{D}} \ \wedge \ \boldsymbol{\sigma}_n(\boldsymbol{u}) = \boldsymbol{g}_{\mathrm{N}} \text{ on } \partial\Omega_{\mathrm{N}}, \tag{6.4}$$

$$u_n \leq 0 \ \wedge \ \sigma_n(\boldsymbol{u}) \leq 0 \ \wedge \ \sigma_n(\boldsymbol{u})\, u_n = 0 \quad \text{on } \partial\Omega_{\mathrm{C}}, \tag{6.5}$$

$$|\boldsymbol{\sigma}_t(\boldsymbol{u})| \leq s \text{ if } \boldsymbol{u}_t = \boldsymbol{0} \ \wedge \ \boldsymbol{\sigma}_t(\boldsymbol{u}) = -s\frac{\boldsymbol{u}_t}{|\boldsymbol{u}_t|} \text{ if } |\boldsymbol{u}_t| > 0 \quad \text{on } \partial\Omega_{\mathrm{C}}, \tag{6.6}$$

where (6.5) are called unilateral contact conditions and (6.6) Tresca friction conditions. The first condition in (6.5) expresses non-interpenetration, whereas the last condition, called complementarity condition, means that either there is contact ($u_n = 0$) or there is no normal force ($\sigma_n(\boldsymbol{u}) = 0$). In (6.6), $s \geq 0$ is a given threshold parameter (more generally, s can be a nonnegative function on $\partial\Omega_{\mathrm{C}}$), and $|\cdot|$ stands for the Euclidean norm in $\mathbb{R}^d$ (or the absolute value depending on the context). The conditions in (6.6) mean that sliding cannot occur as long as the magnitude of the tangential stress $|\boldsymbol{\sigma}_t(\boldsymbol{u})|$ is lower than the threshold s. When the threshold is reached, sliding can happen, in a direction opposite to $\boldsymbol{\sigma}_t(\boldsymbol{u})$ (see, e.g., [113, Chap. 10]). The case of frictionless contact is recovered by setting $s := 0$ in (6.6).

Let us briefly discuss some variants of the above model. On the one hand, bilateral contact with Tresca friction can be considered by keeping (6.6), whereas (6.5) is substituted by the condition

$$u_n = 0. \tag{6.7}$$

This setting is relevant to model persistent contact. In the case of unilateral contact, nonzero tangential stress ($|\boldsymbol{\sigma}_t(\boldsymbol{u})| > 0$) can occur in regions with no-adhesion ($u_n < 0$), which is not expected physically. The setting of bilateral contact prevents such situations. Indeed, since $u_n = 0$, there are no regions with no-adhesion. On the other hand, substituting (6.6) by

$$|\boldsymbol{\sigma}_t(\boldsymbol{u})| \leq \ F|\sigma_n(\boldsymbol{u})| \text{ if } \boldsymbol{u}_t = \boldsymbol{0} \ \wedge \ \boldsymbol{\sigma}_t(\boldsymbol{u}) = -F|\sigma_n(\boldsymbol{u})|\frac{\boldsymbol{u}_t}{|\boldsymbol{u}_t|} \text{ if } |\boldsymbol{u}_t| > 0, \tag{6.8}$$

where $F \geq 0$ is a given friction coefficient, leads to static Coulomb friction. The condition (6.8) is an adaptation of the quasi-static (or dynamic) Coulomb's law, in which the tangential velocity $\dot{\boldsymbol{u}}_t$ plays the same role as the displacement $\boldsymbol{u}_t$. In the rest of this chapter, we focus on the Tresca friction model. This choice is motivated more by mathematical simplicity than physical reasons. Moreover, the Tresca friction model can be useful when Coulomb friction is approximated iteratively.

Recalling the notation $\boldsymbol{H}^1(\Omega) := H^1(\Omega; \mathbb{R}^d)$, we introduce the Hilbert space $\boldsymbol{V}_{\mathrm{D}}$ and the convex cone $\boldsymbol{K}$ such that

$$\boldsymbol{V}_{\mathrm{D}} := \left\{\boldsymbol{v} \in \boldsymbol{H}^1(\Omega) \mid \boldsymbol{v}_{|\partial\Omega_{\mathrm{D}}} = \boldsymbol{0}\right\}, \quad \boldsymbol{K} := \{\boldsymbol{v} \in \boldsymbol{V}_{\mathrm{D}} \mid v_n \leq 0 \text{ on } \partial\Omega_{\mathrm{C}}\},$$

i.e., the Dirichlet condition on $\partial\Omega_{\mathrm{D}}$ is explicitly enforced in the space $\boldsymbol{V}_{\mathrm{D}}$ and the non-interpenetration condition on $\partial\Omega_{\mathrm{C}}$ is explicitly enforced in the cone $\boldsymbol{K}$. We define the following bilinear form and the following linear and nonlinear forms:

$$a(\boldsymbol{v}, \boldsymbol{w}) := (\boldsymbol{\sigma}(\boldsymbol{\varepsilon}(\boldsymbol{v})), \boldsymbol{\varepsilon}(\boldsymbol{w}))_{L^2(\Omega)} = 2\mu(\boldsymbol{\varepsilon}(\boldsymbol{v}), \boldsymbol{\varepsilon}(\boldsymbol{w}))_{L^2(\Omega)} + \lambda(\nabla\cdot\boldsymbol{v}, \nabla\cdot\boldsymbol{w})_{L^2(\Omega)}, \tag{6.9}$$

$$\ell(\boldsymbol{w}) := (\boldsymbol{f}, \boldsymbol{w})_{L^2(\Omega)} + (\boldsymbol{g}_{\mathrm{N}}, \boldsymbol{w})_{L^2(\partial\Omega_{\mathrm{N}})}, \qquad j(\boldsymbol{w}) := \int_{\partial\Omega_{\mathrm{C}}} s|\boldsymbol{w}_t|\,\mathrm{d}s, \tag{6.10}$$

for all $\boldsymbol{v}, \boldsymbol{w} \in \boldsymbol{V}_{\mathrm{D}}$. The weak formulation of (6.3)–(6.6) leads to the following variational inequality:

$$\begin{cases} \text{Find } \boldsymbol{u} \in \boldsymbol{K} \text{ such that} \\ a(\boldsymbol{u}, \boldsymbol{w} - \boldsymbol{u}) + j(\boldsymbol{w}) - j(\boldsymbol{u}) \geq \ell(\boldsymbol{w} - \boldsymbol{u}), \quad \forall \boldsymbol{w} \in \boldsymbol{K}. \end{cases} \tag{6.11}$$

This problem admits a unique solution; see, e.g., [113, Theorem 10.2]. Moreover, this solution is the unique minimizer in $\boldsymbol{K}$ of the energy functional $\mathfrak{E} : \boldsymbol{V}_{\mathrm{D}} \to \mathbb{R}$ such that

$$\mathfrak{E}(\boldsymbol{v}) := \frac{1}{2}a(\boldsymbol{v}, \boldsymbol{v}) + j(\boldsymbol{v}) - \ell(\boldsymbol{v}). \tag{6.12}$$

6.2 HHO-Nitsche Method

The HHO-Nitsche method presented in this section to approximate the model problem (6.11) is inspired by the FEM-Nitsche method devised in [52, 54]. Therefore, we first start with a brief description of the ideas underlying this latter method.

6.2.1 FEM-Nitsche Method

The two keys ideas in the FEM-Nitsche method are on the one hand a reformulation due to [68] of the conditions (6.5)–(6.6) as nonlinear equations and on the other hand the use of a consistent boundary-penalty method inspired by Nitsche [128] to enforce these conditions in the discrete problem.

For all $x \in \mathbb{R}$, let $[x]_{\mathbb{R}^-} := \min(x, 0)$ denote its projection onto $\mathbb{R}^- := (-\infty, 0]$, and for all $\boldsymbol{x} \in \mathbb{R}^d$, let $[\boldsymbol{x}]_\alpha := \boldsymbol{x}$ if $|\boldsymbol{x}| \le \alpha$ and $[\boldsymbol{x}]_\alpha := \alpha \frac{\boldsymbol{x}}{|\boldsymbol{x}|}$ if $|\boldsymbol{x}| > \alpha$ denote its projection onto the closed ball $\mathbb{B}(\boldsymbol{0}, \alpha)$ centered at $\boldsymbol{0}$ and of radius $\alpha > 0$. Let Υ_n and Υ_t be positive functions on $\partial\Omega_{\mathrm{C}}$. Then, as pointed out in [68] (see also [52]), the conditions (6.5)–(6.6) are equivalent to the following statements:

$$\sigma_n(\boldsymbol{u}) = [\tau_n(\boldsymbol{u})]_{\mathbb{R}^-}, \qquad \tau_n(\boldsymbol{u}) := \sigma_n(\boldsymbol{u}) - \Upsilon_n u_n, \tag{6.13}$$

$$\boldsymbol{\sigma}_t(\boldsymbol{u}) = [\boldsymbol{\tau}_t(\boldsymbol{u})]_s, \qquad \boldsymbol{\tau}_t(\boldsymbol{u}) := \boldsymbol{\sigma}_t(\boldsymbol{u}) - \Upsilon_t \boldsymbol{u}_t. \tag{6.14}$$

Let $\mathcal{T}$ be a simplicial mesh of Ω. We assume that Ω is a polyhedron so that the mesh covers Ω exactly, and that every mesh boundary face belongs either to $\partial\Omega_{\mathrm{D}}$, $\partial\Omega_{\mathrm{N}}$, or $\partial\Omega_{\mathrm{C}}$. The corresponding subsets of $\mathcal{F}^\partial$ are denoted by $\mathcal{F}^\partial_{\mathrm{D}}$, $\mathcal{F}^\partial_{\mathrm{N}}$, and $\mathcal{F}^\partial_{\mathrm{C}}$. Let $\mathcal{T}^{\mathrm{C}} \subset \mathcal{T}$ be the collection of the mesh cells having at least one boundary face on $\partial\Omega_{\mathrm{C}}$ and set $\partial T^{\mathrm{C}} := \partial T \cap \partial\Omega_{\mathrm{C}}$ for all $T \in \mathcal{T}^{\mathrm{C}}$. In what follows, we need the following discrete trace inequality which is a slight variant of Lemma 2.4 specialized to $\mathcal{T}^{\mathrm{C}}$: There is C_{dt} such that for all $T \in \mathcal{T}^{\mathrm{C}}$ and all $\boldsymbol{q} \in \mathbb{P}^k_d(T; \mathbb{R}^q)$, $q \in \{1, d\}$,

$$\|\boldsymbol{q}\|_{\boldsymbol{L}^2(\partial T^{\mathrm{C}})} \le C_{\mathrm{dt}} h_T^{-\frac{1}{2}} \|\boldsymbol{q}\|_{\boldsymbol{L}^2(T)}. \tag{6.15}$$

For the time being, we consider an H^1-conforming finite element subspace $\boldsymbol{V}_{h,\mathrm{D}} \subset \boldsymbol{V}_{\mathrm{D}}$. Then, as shown in [52, 54], the FEM-Nitsche method leads to the discrete semilinear form $a_h^{\mathrm{FEM}} : \boldsymbol{V}_{h,\mathrm{D}} \times \boldsymbol{V}_{h,\mathrm{D}} \to \mathbb{R}$ such that $a_h^{\mathrm{FEM}}(\cdot;\cdot) := a(\cdot,\cdot) + n_h^{\mathrm{FEM}}(\cdot;\cdot)$ with

$$\begin{aligned} &n_h^{\mathrm{FEM}}(\boldsymbol{v}_h; \boldsymbol{w}_h) := \\ &-\theta(\Upsilon_n^{-1}\sigma_n(\boldsymbol{v}_h), \sigma_n(\boldsymbol{w}_h))_{L^2(\partial\Omega_{\mathrm{C}})} + (\Upsilon_n^{-1}[\tau_n(\boldsymbol{v}_h)]_{\mathbb{R}^-}, (\tau_n + (\theta - 1)\sigma_n)(\boldsymbol{w}_h))_{L^2(\partial\Omega_{\mathrm{C}})} \\ &-\theta(\Upsilon_t^{-1}\boldsymbol{\sigma}_t(\boldsymbol{v}_h), \boldsymbol{\sigma}_t(\boldsymbol{w}_h))_{\boldsymbol{L}^2(\partial\Omega_{\mathrm{C}})} + (\Upsilon_t^{-1}[\boldsymbol{\tau}_t(\boldsymbol{v}_h)]_s, (\boldsymbol{\tau}_t + (\theta - 1)\boldsymbol{\sigma}_t)(\boldsymbol{w}_h))_{\boldsymbol{L}^2(\partial\Omega_{\mathrm{C}})}, \end{aligned} \tag{6.16}$$

$\boldsymbol{\sigma}(\boldsymbol{v}_h) := \boldsymbol{\sigma}(\boldsymbol{\varepsilon}(\boldsymbol{v}_h))$, $\boldsymbol{\sigma}(\boldsymbol{w}_h) := \boldsymbol{\sigma}(\boldsymbol{\varepsilon}(\boldsymbol{w}_h))$, $\tau_n(\boldsymbol{v}_h)$ and $\boldsymbol{\tau}_t(\boldsymbol{v}_h)$ defined as in (6.13)–(6.14), and $\theta \in \{1, 0, -1\}$ is a symmetry parameter. Choosing $\theta := 1$ leads to a symmetric formulation with a variational structure, choosing $\theta := 0$ is interesting to simplify the implementation by avoiding some terms in the formulation, and choosing $\theta := -1$ allows one to improve on the stability of the method by exploiting its skew-symmetry (see (6.17) where the lower bound vanishes for $\theta = -1$).

The discrete semilinear form a_h^{FEM} enjoys two key properties: (conditional) monotonicity and consistency. On the one hand, monotonicity holds true under a minimal condition on the penalty parameters. We assume that Υ_n and Υ_t are piecewise constant on $\partial\Omega_{\mathrm{C}}$ with $\Upsilon_{n|F} := \gamma_n h_{T_-}^{-1}$ and $\Upsilon_{t|F} := \gamma_t h_{T_-}^{-1}$ with positive parameters γ_n and γ_t, for all $F := \partial T_- \cap \partial\Omega_{\mathrm{C}} \in \mathcal{F}^\partial_{\mathrm{C}}$. Then, assuming that

$$\min(\varrho^{-1}\gamma_n, 2\gamma_t) \ge 3(\theta + 1)^2 C_{\mathrm{dt}}^2 \mu, \tag{6.17}$$

with $\varrho := \max(1, \frac{\lambda}{2\mu})$ and C_{dt} from (6.15), we have (see Lemma 6.2 for the arguments of the proof)

$$a_h^{\mathrm{FEM}}(\boldsymbol{v}_h; \boldsymbol{\delta}_h) - a_h^{\mathrm{FEM}}(\boldsymbol{w}_h; \boldsymbol{\delta}_h) \geq \frac{1}{3} a(\boldsymbol{\delta}_h, \boldsymbol{\delta}_h), \tag{6.18}$$

for all $\boldsymbol{v}_h, \boldsymbol{w}_h \in \boldsymbol{V}_{h,\mathrm{D}}$ with $\boldsymbol{\delta}_h := \boldsymbol{v}_h - \boldsymbol{w}_h$. Concerning consistency, the key observation is that assuming that the exact solution satisfies $\boldsymbol{u} \in \boldsymbol{H}^{1+r}(\Omega)$, $r > \frac{1}{2}$, we have $a_h^{\mathrm{FEM}}(\boldsymbol{u}; \boldsymbol{w}_h) = \ell(\boldsymbol{w}_h)$ for all $\boldsymbol{w}_h \in \boldsymbol{V}_{h,\mathrm{D}}$. Indeed, integration by parts gives $a(\boldsymbol{u}, \boldsymbol{w}_h) - \ell(\boldsymbol{w}_h) = (\boldsymbol{\sigma}_n(\boldsymbol{u}), \boldsymbol{w}_h)_{\boldsymbol{L}^2(\partial\Omega_{\mathrm{C}})}$, while (6.13)–(6.14) imply that

$$n_h^{\mathrm{FEM}}(\boldsymbol{u}; \boldsymbol{w}_h) = (\sigma_n(\boldsymbol{u}), w_{h,n})_{L^2(\partial\Omega_{\mathrm{C}})} + (\boldsymbol{\sigma}_t(\boldsymbol{u}), \boldsymbol{w}_{h,t})_{\boldsymbol{L}^2(\partial\Omega_{\mathrm{C}})} = (\boldsymbol{\sigma}_n(\boldsymbol{u}), \boldsymbol{w}_h)_{\boldsymbol{L}^2(\partial\Omega_{\mathrm{C}})},$$

since $(\tau_n - \sigma_n)(\boldsymbol{w}_h) = \Upsilon_n w_{h,n}$, $(\boldsymbol{\tau}_t - \boldsymbol{\sigma}_t)(\boldsymbol{w}_h) = \Upsilon_t \boldsymbol{w}_{h,t}$, $\boldsymbol{w}_h = w_{h,n}\boldsymbol{n} + \boldsymbol{w}_{h,t}$, and $\boldsymbol{\sigma}_n(\boldsymbol{u}) \cdot \boldsymbol{w}_h = \sigma_n(\boldsymbol{u}) w_{h,n} + \boldsymbol{\sigma}_t(\boldsymbol{u}) \cdot \boldsymbol{w}_{h,t}$.

6.2.2 *Discrete Setting for HHO-Nitsche*

The discrete setting for the HHO-Nitsche method is the same as for the linear elasticity problem in Sect. 4.2.1. As for FEM-Nitsche, we assume that every mesh boundary face belongs either to $\partial\Omega_{\mathrm{D}}$, $\partial\Omega_{\mathrm{N}}$, or $\partial\Omega_{\mathrm{C}}$, and the corresponding subsets of $\mathcal{F}^{\partial}$ are again denoted by $\mathcal{F}^{\partial}_{\mathrm{D}}, \mathcal{F}^{\partial}_{\mathrm{N}}$ and $\mathcal{F}^{\partial}_{\mathrm{C}}$. The HHO-Nitsche method uses the same key ideas as FEM-Nitsche: the nonlinear reformulation (6.13)–(6.14) of the contact and friction conditions, and the weak enforcement of these nonlinear conditions by means of a consistent boundary-penalty method inspired by Nitsche and originally developed in the context of HHO methods in [44].

Our starting point is the equal-order HHO method devised for the linear elasticity problem, where the discrete unknowns are polynomials of degree at most $k \geq 1$ attached to the mesh cells and to the mesh faces. One modification is that the degree of the face unknowns is raised to $(k+1)$ on the boundary faces in $\mathcal{F}^{\partial}_{\mathrm{C}}$. This choice is motivated by the fact that these face unknowns are used to evaluate the quantities τ_n and $\boldsymbol{\tau}_t$ in Nitsche's formulation, so that the error estimate depends on how well these unknowns approximate the trace of the exact solution $\boldsymbol{u}$ on $\partial\Omega_{\mathrm{C}}$. At the same time, this choice increases only marginally the computational costs. For every mesh cell $T \in \mathcal{T}$, let $\mathcal{F}_T$ be the collection of the mesh faces that are subsets of ∂T, which we partition as $\mathcal{F}_T = \mathcal{F}_T^{\mathrm{C}} \cup \mathcal{F}_T^{\setminus\mathrm{C}}$ with $\mathcal{F}_T^{\mathrm{C}} := \mathcal{F}_T \cap \mathcal{F}^{\partial}_{\mathrm{C}}$ (the subset $\mathcal{F}_T^{\mathrm{C}}$ is empty for all $T \notin \mathcal{T}^{\mathrm{C}}$). Then, the local HHO discrete space is

$$\hat{\boldsymbol{V}}_T^k := \mathbb{P}_d^k(T) \times \mathbb{P}_{d-1}^{k+}(\mathcal{F}_T), \quad \mathbb{P}_{d-1}^{k+}(\mathcal{F}_T) := \bigtimes_{F \in \mathcal{F}_T^{\setminus\mathrm{C}}} \mathbb{P}_{d-1}^k(F) \times \bigtimes_{F \in \mathcal{F}_T^{\mathrm{C}}} \mathbb{P}_{d-1}^{k+1}(F). \tag{6.19}$$

A generic element in $\hat{\boldsymbol{V}}_T^k$ is denoted by $\hat{\boldsymbol{v}}_T := (\boldsymbol{v}_T, \boldsymbol{v}_{\partial T})$. The discrete unknowns are illustrated in Fig. 6.1.

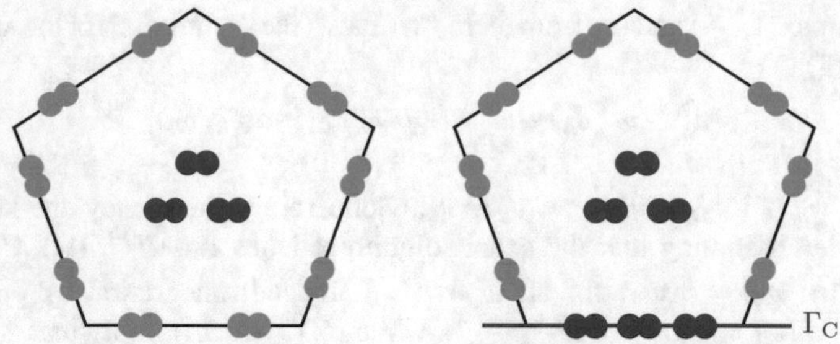

(a) Pentagonal cell with no contact face ($\mathcal{F}_T^{\setminus C} = \mathcal{F}_T$)

(b) Pentagonal cell with one contact face in red ($\mathcal{F}_T^{\setminus C} \subsetneq \mathcal{F}_T$)

Fig. 6.1 Face (red or violet) and cell (blue) unknowns in $\hat{V}_T^k$ for $k = 1$ and $d = 2$ (each dot represents a basis function)

We consider as in Chap. 5 the local strain reconstruction operator $\boldsymbol{E}_T : \hat{\boldsymbol{V}}_T^k \to \mathbb{P}_d^k(T; \mathbb{R}_{\text{sym}}^{d\times d})$ such that for all $\hat{\boldsymbol{v}}_T \in \hat{\boldsymbol{V}}_T^k$,

$$(\boldsymbol{E}_T(\hat{\boldsymbol{v}}_T), \boldsymbol{q})_{\boldsymbol{L}^2(T)} = -(\boldsymbol{v}_T, \nabla{\cdot}\boldsymbol{q})_{\boldsymbol{L}^2(T)} + (\boldsymbol{v}_{\partial T}, \boldsymbol{q}\boldsymbol{n}_T)_{\boldsymbol{L}^2(\partial T)}, \tag{6.20}$$

for all $\boldsymbol{q} \in \mathbb{P}_d^k(T; \mathbb{R}_{\text{sym}}^{d\times d})$. The local discrete divergence operator $D_T : \hat{\boldsymbol{V}}_T^k \to \mathbb{P}_d^k(T)$ is simply defined by taking the trace of the reconstructed strain tensor, i.e., for all $\hat{\boldsymbol{v}}_T \in \hat{\boldsymbol{V}}_T^k$, we set $D_T(\hat{\boldsymbol{v}}_T) := \mathrm{tr}(\boldsymbol{E}_T(\hat{\boldsymbol{v}}_T))$. The local stabilization operator $\boldsymbol{S}_{\partial T} : \hat{\boldsymbol{V}}_T^k \to \mathbb{P}_{d-1}^{k+}(\mathcal{F}_T)$ is readily adapted from the one considered for linear elasticity by setting for all $\hat{\boldsymbol{v}}_T \in \hat{\boldsymbol{V}}_T^k$,

$$\boldsymbol{S}_{\partial T}(\hat{\boldsymbol{v}}_T) := \boldsymbol{\Pi}_{\partial T}^{k+}\Big(\boldsymbol{v}_{T|\partial T} - \boldsymbol{v}_{\partial T} + \big((I - \boldsymbol{\Pi}_T^k)\boldsymbol{U}_T(\hat{\boldsymbol{v}}_T)\big)_{|\partial T}\Big), \tag{6.21}$$

where $\boldsymbol{\Pi}_{\partial T}^{k+}$ is the L^2-orthogonal projections onto $\mathbb{P}_{d-1}^{k+}(\mathcal{F}_T)$ and the displacement reconstruction operator $\boldsymbol{U}_T : \hat{\boldsymbol{V}}_T^k \to \mathbb{P}_d^{k+1}(T)$ is defined in (4.26). Using the above operators leads to the following local bilinear form defined on $\hat{\boldsymbol{V}}_T^k \times \hat{\boldsymbol{V}}_T^k$:

$$\begin{aligned} a_T(\hat{\boldsymbol{v}}_T, \hat{\boldsymbol{w}}_T) &:= 2\mu(\boldsymbol{E}_T(\hat{\boldsymbol{v}}_T), \boldsymbol{E}_T(\hat{\boldsymbol{w}}_T))_{\boldsymbol{L}^2(T)} + \lambda(D_T(\hat{\boldsymbol{v}}_T), D_T(\hat{\boldsymbol{w}}_T))_{L^2(T)} \\ &\quad + 2\mu h_T^{-1}(\boldsymbol{S}_{\partial T}(\hat{\boldsymbol{v}}_T), \boldsymbol{S}_{\partial T}(\hat{\boldsymbol{w}}_T))_{\boldsymbol{L}^2(\partial T)}. \end{aligned} \tag{6.22}$$

For simplicity, we employ the Nitsche technique only on the subset $\partial\Omega_C$ where the nonlinear frictional contact conditions are enforced, whereas we resort to a strong enforcement of the homogeneous Dirichlet condition on the subset $\partial\Omega_D$. The global discrete spaces for the HHO-Nitsche method are

$$\hat{V}_h^k := \bigtimes_{T\in\mathcal{T}} \mathbb{P}_d^k(T) \times \bigtimes_{F\in\mathcal{F}^\circ\cup\mathcal{F}_N^\partial\cup\mathcal{F}_D^\partial} \mathbb{P}_{d-1}^k(F) \times \bigtimes_{F\in\mathcal{F}_C^\partial} \mathbb{P}_{d-1}^{k+1}(F). \tag{6.23}$$

$$\hat{V}_{h,D}^k := \{\hat{v}_h \in \hat{V}_h^k \mid v_F = \mathbf{0},\ \forall F \in \mathcal{F}_D^\partial\}, \tag{6.24}$$

leading to the notation $\hat{v}_h := \big((v_T)_{T\in\mathcal{T}}, (v_F)_{F\in\mathcal{F}}\big)$ for a generic element $\hat{v}_h \in \hat{V}_h^k$. For all $T \in \mathcal{T}$, we denote by $\hat{v}_T := (v_T, v_{\partial T} := (v_F)_{F\in\mathcal{F}_T}) \in \hat{V}_T^k$ the local components of $\hat{v}_h$ attached to the mesh cell T and the faces composing ∂T, and for any mesh face $F \in \mathcal{F}$, we denote by v_F the component of $\hat{v}_h$ attached to the face F. The global discrete bilinear form related to the linear elasticity part of the problem is, as usual, assembled cellwise by setting $a_h(\hat{v}_h, \hat{w}_h) := \sum_{T\in\mathcal{T}} a_T(\hat{v}_T, \hat{w}_T)$, and it remains to extend to the HHO setting the Nitsche-like semilinear form n_h^{FEM} defined in (6.16). To this purpose, we set for all $T \in \mathcal{T}^{\mathrm{C}}$ and all $\hat{w}_T \in \hat{V}_T^k$,

$$\sigma(\hat{w}_T) := 2\mu E_T(\hat{w}_T) + \lambda D_T(\hat{w}_T) I_d \in \mathbb{P}_d^k(T; \mathbb{R}_{\mathrm{sym}}^{d\times d}), \tag{6.25}$$

with the decomposition $\sigma(\hat{w}_T)n_T := \sigma_n(\hat{w}_T)n_T + \sigma_t(\hat{w}_T)$. Inspired by (6.13)–(6.14), we also introduce the linear operators $\tau_n : \hat{V}_T^k \to \mathbb{P}_{d-1}^{k+1}(\mathcal{F}_T^{\mathrm{C}})$ and $\tau_t : \hat{V}_T^k \to \mathbb{P}_{d-1}^{k+1}(\mathcal{F}_T^{\mathrm{C}})$ such that (notice the use of the face component on the right-hand side)

$$\tau_n(\hat{w}_T) := \sigma_n(\hat{w}_T) - \Upsilon_n w_{\partial T,n}, \qquad \tau_t(\hat{w}_T) := \sigma_t(\hat{w}_T) - \Upsilon_t w_{\partial T,t}, \tag{6.26}$$

together with the decomposition $w_{\partial T} := w_{\partial T,n} n_T + w_{\partial T,t}$ for the face component. We then set $n_h^{\mathrm{HHO}}(\hat{v}_h, \hat{w}_h) := \sum_{T\in\mathcal{T}^{\mathrm{C}}} n_T^{\mathrm{HHO}}(\hat{v}_T, \hat{w}_T)$ for all $\hat{v}_h, \hat{w}_h \in \hat{V}_h^k$ with

$$\begin{aligned} n_T^{\mathrm{HHO}}(\hat{v}_T, \hat{w}_T) &:= \\ &- \theta(\Upsilon_n^{-1}\sigma_n(\hat{v}_T), \sigma_n(\hat{w}_T))_{L^2(\partial T^{\mathrm{C}})} + (\Upsilon_n^{-1}[\tau_n(\hat{v}_T)]_{\mathbb{R}^-}, (\tau_n + (\theta-1)\sigma_n)(\hat{w}_T))_{L^2(\partial T^{\mathrm{C}})} \\ &- \theta(\Upsilon_t^{-1}\sigma_t(\hat{v}_T), \sigma_t(\hat{w}_T))_{L^2(\partial T^{\mathrm{C}})} + (\Upsilon_t^{-1}[\tau_t(\hat{v}_T)]_s)((\tau_t + (\theta-1)\sigma_t)(\hat{w}_T))_{L^2(\partial T^{\mathrm{C}})}, \end{aligned} \tag{6.27}$$

where $\theta \in \{1, 0, -1\}$ is again the symmetry parameter. This leads to the following discrete HHO-Nitsche problem:

$$\begin{cases} \text{Find } \hat{u}_h \in \hat{V}_{h,D}^k \text{ such that} \\ a_h^{\mathrm{HHO}}(\hat{u}_h; \hat{w}_h) = \ell_h(\hat{w}_h) \quad \forall \hat{w}_h \in \hat{V}_{h,D}^k, \end{cases} \tag{6.28}$$

with $a_h^{\mathrm{HHO}}(\cdot;\cdot) := a_h(\cdot,\cdot) + n_h^{\mathrm{HHO}}(\cdot;\cdot)$ and the linear form on the right-hand side is defined as $\ell_h(\hat{w}_h) := (f, w_{\mathcal{T}})_{L^2(\Omega)} + (g_{\mathrm{N}}, w_{\mathcal{F}})_{L^2(\partial\Omega_{\mathrm{N}})}$.

Remark 6.1 (*Literature*) The above HHO-Nitsche method for contact and friction problems is devised and analyzed in [53]. This is, to our knowledge, so far the only discretization method supporting polyhedral meshes that benefits from the same features as the FEM-Nitsche method devised in [52, 54], namely optimal error estimates without additional assumptions on the contact/friction set (see also [55] for the analysis of FEM-Nitsche). Notice also that [53] tracks the dependency of the penalty parameters and error estimates on the Lamé parameters μ and λ. Other polyhedral discretization methods for contact/friction problems, that however do not hinge on Nitsche's approach, include virtual element [147, 152], weak Galerkin [101], and hybridizable discontinuous Galerkin [155] methods. □

6.2.3 Stability and Error Analysis

In this section we outline the stability and error analysis for the above HHO-Nitsche method, and we refer the reader to [53] for more details.

Lemma 6.2 (Monotonicity, well-posedness) *Assume that Υ_n and Υ_t are piecewise constant on $\partial\Omega_C$ with $\Upsilon_{n|F} := \gamma_n h_{T_-}^{-1}$ and $\Upsilon_{t|F} := \gamma_t h_{T_-}^{-1}$ with positive parameters γ_n and γ_t, for all $F := \partial T_- \cap \partial\Omega_C \in \mathcal{F}_C^\partial$. Then, assuming that the minimality condition* (6.17) *on γ_n and γ_t holds true, we have*

$$a_h^{\mathrm{HHO}}(\hat{\boldsymbol{v}}_h; \hat{\boldsymbol{\delta}}_h) - a_h^{\mathrm{HHO}}(\hat{\boldsymbol{w}}_h; \hat{\boldsymbol{\delta}}_h) \geq \frac{1}{3} a_h(\hat{\boldsymbol{\delta}}_h, \hat{\boldsymbol{\delta}}_h), \tag{6.29}$$

for all $\hat{\boldsymbol{v}}_h, \hat{\boldsymbol{w}}_h \in \hat{\boldsymbol{V}}_{h,D}^k$ with $\hat{\boldsymbol{\delta}}_h := \hat{\boldsymbol{v}}_h - \hat{\boldsymbol{w}}_h$. Moreover, the discrete problem (6.28) *is well-posed.*

Proof (i) We have $n_h^{\mathrm{HHO}}(\hat{\boldsymbol{v}}_h; \hat{\boldsymbol{\delta}}_h) - n_h^{\mathrm{HHO}}(\hat{\boldsymbol{w}}_h; \hat{\boldsymbol{\delta}}_h) = -\sum_{T\in\mathcal{T}^{\mathrm{C}}}(A_{T,n} + A_{T,t})$ with

$$\begin{aligned}
\frac{\gamma_n}{h_T} A_{T,n} &:= \theta \|\sigma_n(\hat{\boldsymbol{\delta}}_T)\|^2_{L^2(\partial T^{\mathrm{C}})} - (\delta\tau_n, \tau_n(\hat{\boldsymbol{\delta}}_T))_{L^2(\partial T^{\mathrm{C}})} - (\theta - 1)(\delta\tau_n, \sigma_n(\hat{\boldsymbol{\delta}}_T))_{L^2(\partial T^{\mathrm{C}})},\\
\frac{\gamma_t}{h_T} A_{T,t} &:= \theta \|\boldsymbol{\sigma}_t(\hat{\boldsymbol{\delta}}_T)\|^2_{\boldsymbol{L}^2(\partial T^{\mathrm{C}})} - (\delta\boldsymbol{\tau}_t, \boldsymbol{\tau}_t(\hat{\boldsymbol{\delta}}_T))_{\boldsymbol{L}^2(\partial T^{\mathrm{C}})} - (\theta - 1)(\delta\boldsymbol{\tau}_t, \boldsymbol{\sigma}_t(\hat{\boldsymbol{\delta}}_T))_{\boldsymbol{L}^2(\partial T^{\mathrm{C}})},
\end{aligned}$$

with $\delta\tau_n := \big[\tau_n(\hat{\boldsymbol{v}}_T)\big]_{\mathbb{R}^-} - \big[\tau_n(\hat{\boldsymbol{w}}_T)\big]_{\mathbb{R}^-}$ and $\delta\boldsymbol{\tau}_t := \big[\boldsymbol{\tau}_t(\hat{\boldsymbol{v}}_T)\big]_s - \big[\boldsymbol{\tau}_t(\hat{\boldsymbol{w}}_T)\big]_s$. Using that $([x]_{\mathbb{R}^-} - [y]_{\mathbb{R}^-})(x-y) \geq ([x]_{\mathbb{R}^-} - [y]_{\mathbb{R}^-})^2 \geq 0$ for all $x, y \in \mathbb{R}$, Young's inequality and the identity $\theta + \frac{1}{4}(\theta-1)^2 = \frac{1}{4}(\theta+1)^2$ shows that

$$A_{T,n} \leq \frac{1}{4}(\theta+1)^2 \|\sigma_n(\hat{\boldsymbol{\delta}}_T)\|^2_{L^2(\partial T^{\mathrm{C}})} \leq \frac{1}{4}(\theta+1)^2 \frac{C_{\mathrm{dt}}^2}{\gamma_n} \|\sigma_n(\hat{\boldsymbol{\delta}}_T)\|^2_{L^2(T)},$$

where the last bound follows from the discrete trace inequality (6.15). Using the definition (6.25) of the discrete stress, the triangle and Young's inequalities gives

$$A_{T,n} \leq (\theta+1)^2 \frac{C_{\mathrm{dt}}^2}{\gamma_n} \mu_{\varrho} \times \big(2\mu \|\boldsymbol{E}_T(\hat{\boldsymbol{\delta}}_T)\|^2_{\mathbb{L}^2(T)} + \lambda \|D_T(\hat{\boldsymbol{\delta}}_T)\|^2_{L^2(T)}\big),$$

recalling that $\varrho := \max(1, \frac{\lambda}{2\mu})$. Using similar arguments, and in particular that $([\boldsymbol{x}]_s - [\boldsymbol{y}]_s)\cdot(\boldsymbol{x} - \boldsymbol{y}) \geq |[\boldsymbol{x}]_s - [\boldsymbol{y}]_s|^2 \geq 0$ for all $\boldsymbol{x}, \boldsymbol{y} \in \mathbb{R}^d$, shows that

$$A_{T,t} \leq (\theta + 1)^2 \frac{C_{\mathrm{dt}}^2}{\gamma_t} \mu \times 2\mu \|\boldsymbol{E}_T(\hat{\boldsymbol{\delta}}_T)\|_{\boldsymbol{L}^2(T)}^2.$$

Putting these bounds together and using the condition (6.17) on the penalty parameters γ_n and γ_t proves that

$$n_h^{\mathrm{HHO}}(\hat{\boldsymbol{v}}_h; \hat{\boldsymbol{\delta}}_h) - n_h^{\mathrm{HHO}}(\hat{\boldsymbol{w}}_h; \hat{\boldsymbol{\delta}}_h) \geq -\frac{2}{3} \sum_{T \in \mathcal{T}^{\mathrm{C}}} \big(2\mu \|\boldsymbol{E}_T(\hat{\boldsymbol{\delta}}_T)\|_{\boldsymbol{L}^2(T)}^2 + \lambda \|D_T(\hat{\boldsymbol{\delta}}_T)\|_{L^2(T)}^2\big),$$

so that $n_h^{\mathrm{HHO}}(\hat{\boldsymbol{v}}_h; \hat{\boldsymbol{\delta}}_h) - n_h^{\mathrm{HHO}}(\hat{\boldsymbol{w}}_h; \hat{\boldsymbol{\delta}}_h) \geq -\frac{2}{3} a_h(\hat{\boldsymbol{\delta}}_h, \hat{\boldsymbol{\delta}}_h)$. This proves (6.29) since $a_h^{\mathrm{HHO}}(\hat{\boldsymbol{v}}_h; \hat{\boldsymbol{\delta}}_h) - a_h^{\mathrm{HHO}}(\hat{\boldsymbol{w}}_h; \hat{\boldsymbol{\delta}}_h) = a_h(\hat{\boldsymbol{\delta}}_h, \hat{\boldsymbol{\delta}}_h) + n_h^{\mathrm{HHO}}(\hat{\boldsymbol{v}}_h; \hat{\boldsymbol{\delta}}_h) - n_h^{\mathrm{HHO}}(\hat{\boldsymbol{w}}_h; \hat{\boldsymbol{\delta}}_h)$.
(ii) Recalling (5.13) shows that a_h is coercive on $\hat{\boldsymbol{V}}_{h,\mathrm{D}}^k$ with respect to the norm $|\hat{\boldsymbol{v}}_h|_{\boldsymbol{\varepsilon},h}^2 := \sum_{T\in\mathcal{T}} |\hat{\boldsymbol{v}}_T|_{\boldsymbol{\varepsilon},T}^2$ with $|\hat{\boldsymbol{v}}_T|_{\boldsymbol{\varepsilon},T}^2 := \|\boldsymbol{\varepsilon}(\boldsymbol{v}_T)\|_{\mathbb{L}^2(T)}^2 + h_T^{-1}\|\boldsymbol{v}_T - \boldsymbol{v}_{\partial T}\|_{\boldsymbol{L}^2(\partial T)}^2$. Therefore, combining the monotonicity property (6.29) with the arguments from [29, Corollary 15, p. 126] (see also [52]) proves that (6.28) is well-posed. □

Let us finally state without proof an $\boldsymbol{H}^1$-error estimate. Referring to [53] for more details, we observe that the bound on the consistency error combines the arguments from the proof of Lemma 4.8 (for linear elasticity) and the arguments at the end of Sect. 6.2.1 (for FEM-Nitsche). Let $\boldsymbol{E}_{\mathcal{T}}$ and $D_{\mathcal{T}}$ be the global reconstruction operators such that $\boldsymbol{E}_{\mathcal{T}}(\hat{\boldsymbol{v}}_h)_{|T} := \boldsymbol{E}_T(\hat{\boldsymbol{v}}_T)$ and $D_{\mathcal{T}}(\hat{\boldsymbol{v}}_h)_{|T} := D_T(\hat{\boldsymbol{v}}_T)$ for all $T \in \mathcal{T}$ and all $\hat{\boldsymbol{v}}_h \in \hat{\boldsymbol{V}}_{h,\mathrm{D}}^k$. Let $\boldsymbol{\Pi}_{\mathcal{T}}^l$, $l \in \{k, k+1\}$, denote the global $\boldsymbol{L}^2$-orthogonal projection onto the corresponding piecewise polynomial space.

Theorem 6.3 ($\boldsymbol{H}^1$-error estimate) *Assume that the penalty parameters satisfy the tighter condition* $\min(\varrho^{-1}\gamma_n, 2\gamma_t) \geq 3\big((\theta+1)^2 + (4 + (\theta-1)^2)\big)C_{\mathrm{dt}}^2\mu$. *Let* $\hat{\boldsymbol{u}}_h$ *be the discrete solution of (6.28) with local components* $\hat{\boldsymbol{u}}_T$ *for all* $T \in \mathcal{T}$. *Assume that the exact solution satisfies* $\boldsymbol{u} \in \boldsymbol{H}^{1+r}(\Omega)$, $r > \frac{1}{2}$. *There is* C, *uniform with respect to* μ *and* λ, *such that*

$$2\mu\|\boldsymbol{\varepsilon}(\boldsymbol{u}) - \boldsymbol{E}_{\mathcal{T}}(\hat{\boldsymbol{u}}_h)\|_{\mathbb{L}^2(\Omega)}^2 + \lambda\|\nabla\cdot\boldsymbol{u} - D_{\mathcal{T}}(\hat{\boldsymbol{u}}_h)\|_{L^2(\Omega)}^2 \leq C\Psi(\boldsymbol{u}),$$

with $\Psi(\boldsymbol{u}) := \Psi^{\mathrm{EL}}(\boldsymbol{u}) + \Psi^{\mathrm{CO}}(\boldsymbol{u}) + \Psi^{\mathrm{FR}}(\boldsymbol{u})$, $\Psi^{\mathrm{EL}}(\boldsymbol{u}) := 2\mu|\boldsymbol{\varepsilon}(\boldsymbol{u}) - \boldsymbol{\Pi}_{\mathcal{T}}^k(\boldsymbol{\varepsilon}(\boldsymbol{u}))|_{\sharp,\mathcal{T}}^2 + 2\mu\|\boldsymbol{\varepsilon}_{\mathcal{T}}(\boldsymbol{u} - \boldsymbol{\Pi}_{\mathcal{T}}^{k+1}(\boldsymbol{u}))\|_{\mathbb{L}^2(\Omega)}^2 + \lambda\varrho\|\nabla\cdot\boldsymbol{u} - \Pi_{\mathcal{T}}^k(\nabla\cdot\boldsymbol{u})\|_{\dagger,\mathcal{T}}^2$, *the (semi)norms* $|\cdot|_{\sharp,\mathcal{T}}$ *and* $\|\cdot\|_{\dagger,\mathcal{T}}$ *defined in (4.46), and*

$$\Psi^{\mathrm{CO}}(\boldsymbol{u}) := \sum_{T\in\mathcal{T}^{\mathrm{C}}} \Big(\frac{h_T}{\gamma_n}\|\sigma_n(\boldsymbol{u}) - \sigma_n(\hat{\boldsymbol{I}}_T^k(\boldsymbol{u}))\|_{L^2(\partial T^{\mathrm{C}})}^2 + \frac{\gamma_n}{h_T}\|\delta u_{T,n}\|_{L^2(\partial T^{\mathrm{C}})}^2\Big), \tag{6.30}$$

$$\Psi^{\mathrm{FR}}(\boldsymbol{u}) := \sum_{T\in\mathcal{T}^{\mathrm{C}}} \Big(\frac{h_T}{\gamma_t}\|\boldsymbol{\sigma}_t(\boldsymbol{u}) - \boldsymbol{\sigma}_t(\hat{\boldsymbol{I}}_T^k(\boldsymbol{u}))\|_{\boldsymbol{L}^2(\partial T^{\mathrm{C}})}^2 + \frac{\gamma_t}{h_T}\|\delta \boldsymbol{u}_{T,t}\|_{\boldsymbol{L}^2(\partial T^{\mathrm{C}})}^2\Big), \tag{6.31}$$

where the local reduction operator is defined such that $\hat{\boldsymbol{I}}_T^k(\boldsymbol{u}) := (\boldsymbol{\Pi}_T^k(\boldsymbol{u}), \boldsymbol{\Pi}_{\partial T}^{k+}(\boldsymbol{u}_{|\partial T}))$, *and* $(\delta u_{T,n}, \delta \boldsymbol{u}_{T,t})$ *are the normal and tangential components of* $\boldsymbol{u}_{|\partial T} - \boldsymbol{\Pi}_{\partial T}^{k+}(\boldsymbol{u}_{|\partial T})$.

An error estimate on the satisfaction of the contact/friction conditions is also given in [53, Theorem 12]. Moreover, provided the exact solution satisfies $\boldsymbol{u} \in \boldsymbol{H}^{1+r}(\mathcal{T})$ and $\nabla{\cdot}\boldsymbol{u} \in H^r(\mathcal{T})$ with $r \in (\frac{1}{2}, k+1]$, Theorem 6.3 implies that the $\boldsymbol{H}^1$-error decays optimally with rate $O(h^{k+1})$. Notice however that in general, when there is a transition between contact and no-contact, the best expected regularity exponent is $r = \frac{5}{2} - \varepsilon$, $\varepsilon > 0$, so that the maximal convergence rate is $O(h^{\frac{3}{2}-\varepsilon})$ and is reached for $k = 1$. Finally, we notice that using face polynomials of degree $(k+1)$ on the faces in $\mathcal{F}_{\mathrm{C}}^{\partial}$ is crucial to estimate optimally the rightmost terms in (6.30)–(6.31).

Remark 6.4 (*Quasi-incompressible limit*) In this situation, the factor ϱ can be very large. The minimality condition (6.17) is robust with respect to the quasi-incompressible limit in the two following situations: (i) for the skew-symmetric variant $\theta = -1$, since the penalty parameters γ_n and γ_t need only to be positive real numbers (instead, for $\theta \in \{0, 1\}$, this property is lost for γ_n which needs to scale as $\mu\varrho$); (ii) for bilateral contact and any value of θ, since only the parameter γ_t is used and its value remains independent of ϱ. In contrast, the error estimate from Theorem 6.3 is affected by large values of ϱ. The numerical experiments reported in [53] do not indicate, however, any sign of lack of robustness. □

6.3 Numerical Example

We consider a prototype for an industrial application that simulates the installation of a notched plug in a rigid pipe. The mesh is composed of 21,200 hexahedra and 510 prisms (for symmetry reasons, only one quarter of the pipe is discretized). The notched plug has a length of 56 mm and an outer radius of 8 mm. The pipe is supposed to be rigid and has an inner radius of 8.77 mm (there is an initial gap of 0.77 mm between the plug and the pipe). The contact zone with Tresca's friction ($s := 3{,}000$ MPa) is between the rigid pipe and the ten notches of the plug. In the actual industrial setting, an indenter imposes a displacement to the upper surface of the plug. To simplify, sufficiently large vertical and horizontal forces are applied to the upper surface of the plug to impose a contact between the pipe and the notches. The material parameters for the plug are $\mu := 80{,}769$ MPa and $\lambda := 121{,}154$ MPa (which correspond to a Young modulus $E = 210{,}000$ MPa and a Poisson ratio $\nu = 0.3$). The simulation is performed using $k := 1$, the symmetric variant $\theta := 1$, and the penalty parameters $\gamma_n = \gamma_t := 2\mu$). The discrete nonlinear problem (6.28) is solved by a generalized Newton's method as in [68]. The von Mises stress is plotted in Fig. 6.2 on the deformed configuration (a zoom on the contact zone is shown). We remark that there is contact between the notches and the pipe. Finally, the normal stress σ_n

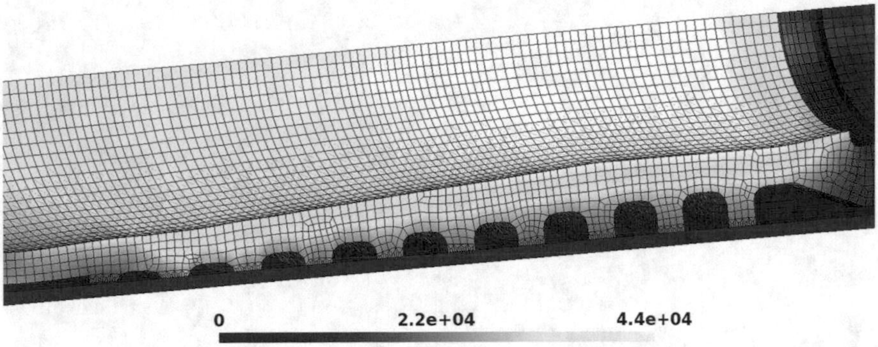

Fig. 6.2 Notch plug (zoom on contact zone): von Mises stress (MPa) on the deformed configuration

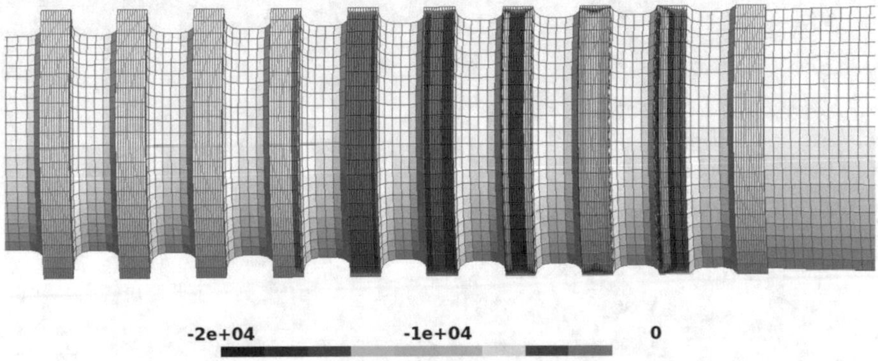

Fig. 6.3 Notch plug: normal stress σ_n (MPa) on the contact zone

is visualized in Fig. 6.3 on the inferior surface of the plug. We remark that all the notches are in contact except the first three (from left to right) and the last one (where $\sigma_n = 0$), and that a transition between contact and non-contact is located at the fourth notch. Moreover, the maximal value of the normal stress is reached at the extremity of the notches.

Chapter 7
Plasticity

Modeling plasticity problems is particularly relevant in nonlinear solid mechanics since plasticity can have a major influence on the behavior of a mechanical structure. One difficulty is that the plastic deformations are generally assumed to be incompressible, leading to volume-locking problems if (low-order) H^1-conforming finite elements are used. Mixed methods avoid these problems, but need additional globally coupled unknowns to enforce the incompressibility of the plastic deformations. Discontinuous Galerkin methods also avoid locking problems, but generally require to perform the integration of the behavior law at quadrature nodes located on the mesh faces, and not only in the mesh cells. In contrast, HHO methods are free of volume locking, only handle primal unknowns, and integrate the behavior law only at quadrature nodes in the mesh cells.

7.1 Plasticity Model

Contrary to the elastic and hyperlastic models, the elastoplastic model is based on the assumption that the deformations are no longer reversible. We place ourselves within the framework of generalized standard materials [102, 118]. Moreover, the plasticity model is assumed to be strain-hardening (or perfect) and rate-independent, i.e., the speed of the deformations has no influence on the solution. For this reason, only the incremental plasticity problem with a pseudo-time is considered.

M. Cicuttin et al., *Hybrid High-Order Methods*,
SpringerBriefs in Mathematics,
https://doi.org/10.1007/978-3-030-81477-9_7

7.1.1 Kinematics and Additive Decomposition

We consider an elastoplastic material body that occupies the domain Ω in the reference configuration. Here, $\Omega \subset \mathbb{R}^d$, $d \in \{2, 3\}$, is a bounded connected Lipschitz domain with unit outward normal $\boldsymbol{n}$ and boundary partitioned as $\partial\Omega = \overline{\partial\Omega_{\mathrm{N}}} \cup \overline{\partial\Omega_{\mathrm{D}}}$ with two relatively open and disjoint subsets $\partial\Omega_{\mathrm{N}}$ and $\partial\Omega_{\mathrm{D}}$. Due to the deformation, a point $\boldsymbol{x} \in \Omega$ is mapped to a point $\boldsymbol{x}'(t) = \boldsymbol{x} + \boldsymbol{u}(t, \boldsymbol{x})$ in the equilibrium configuration, where $\boldsymbol{u} : J \times \Omega \to \mathbb{R}^d$ is the displacement field and J is the pseudo-time interval. The deformation gradient $\boldsymbol{F}(\boldsymbol{u}) = \boldsymbol{I}_d + \nabla\boldsymbol{u}$ takes values in $\mathbb{R}_+^{d\times d}$, which is the set of $\mathbb{R}^{d\times d}$-matrices with positive determinant.

The regimes of infinitesimal and finite deformations are considered here. For infinitesimal deformations, we consider the linearized strain tensor (see Sect. 4.1.1)

$$\boldsymbol{\varepsilon}(\boldsymbol{u}) := \frac{1}{2}(\nabla\boldsymbol{u} + \nabla\boldsymbol{u}^{\mathsf{T}}). \tag{7.1}$$

For finite deformations, we adopt the logarithmic strain framework [123] leading to the following strain tensor:

$$\boldsymbol{E}(\boldsymbol{u}) := \frac{1}{2} \ln\left(\boldsymbol{F}(\boldsymbol{u})^{\mathsf{T}}\boldsymbol{F}(\boldsymbol{u})\right) =: \mathcal{L}(\boldsymbol{F}(\boldsymbol{u})), \tag{7.2}$$

with the transformation $\mathcal{L} : \mathbb{R}_+^{d\times d} \to \mathbb{R}_{\mathrm{sym}}^{d\times d}$ such that $\mathcal{L}(\boldsymbol{t}) := \frac{1}{2}\ln(\boldsymbol{t}^{\mathsf{T}}\boldsymbol{t})$. Evaluating $\mathcal{L}(\boldsymbol{t})$ requires to perform an eigenvalue decomposition of $\boldsymbol{t}^{\mathsf{T}}\boldsymbol{t}$.

Both strain tensors defined in (7.1)–(7.2) are symmetric, and we notice that for infinitesimal deformations where $\|\nabla\boldsymbol{u}\|_{\ell^2} \ll 1$ with $\|\nabla\boldsymbol{u}\|_{\ell^2} := (\nabla\boldsymbol{u} : \nabla\boldsymbol{u})^{\frac{1}{2}}$, we have $\boldsymbol{E}(\boldsymbol{u}) \approx \boldsymbol{\varepsilon}(\boldsymbol{u})$. To avoid the proliferation of cases, we work in this chapter with the tensor $\boldsymbol{E}(\boldsymbol{u})$, keeping in mind that everything can be adapted to infinitesimal deformations.

7.1.2 Helmholtz Free Energy and Yield Function

In the framework of generalized standard materials, the material state is described locally by the strain tensor $\boldsymbol{E} \in \mathbb{R}_{\mathrm{sym}}^{d\times d}$ (we drop the dependency on $\boldsymbol{u}$), the plastic strain tensor $\boldsymbol{E}^{\mathrm{p}} \in \mathbb{R}_{\mathrm{sym}}^{d\times d}$ which is trace-free, and a finite collection of internal variables $\boldsymbol{\alpha} := (\alpha_1, \ldots, \alpha_m) \in \mathbb{R}^m$. The elastic strain tensor is then defined as follows:

$$\boldsymbol{E}^{\mathrm{e}} := \boldsymbol{E} - \boldsymbol{E}^{\mathrm{p}} \in \mathbb{R}_{\mathrm{sym}}^{d\times d}, \qquad \mathrm{tr}(\boldsymbol{E}^{\mathrm{p}}) = 0. \tag{7.3}$$

The Helmholtz free energy $\Psi : \mathbb{R}_{\mathrm{sym}}^{d\times d} \times \mathbb{R}^m \to \mathbb{R}$ acts on a generic pair $(\boldsymbol{e}, \boldsymbol{a})$ representing the elastic strain tensor and the internal variables. We assume that this function satisfies the following hypothesis.

Hypothesis 7.1 (*Helmholtz free energy*) Ψ can be decomposed additively into an elastic and a plastic part as follows:

$$\Psi(\boldsymbol{e}, \boldsymbol{a}) := \frac{1}{2}\boldsymbol{e} : \mathbb{A} : \boldsymbol{e} + \Psi^{\mathrm{p}}(\boldsymbol{a}) \tag{7.4}$$

where $\Psi^{\mathrm{p}} : \mathbb{R}^m \to \mathbb{R}$ is convex (and strongly convex for strain-hardening plasticity), and the elastic modulus is $\mathbb{A} := 2\mu\mathbb{I}^{\mathrm{s}} + \lambda \boldsymbol{I} \otimes \boldsymbol{I}$, with $\mu > 0$, $3\lambda + 2\mu > 0$, $(\mathbb{I}^{\mathrm{s}})_{ij,kl} := \frac{1}{2}(\delta_{ik}\delta_{jl} + \delta_{il}\delta_{jk})$, and $(\boldsymbol{I} \otimes \boldsymbol{I})_{ij,kl} = \delta_{ij}\delta_{kl}$ for all $1 \le i, j, k, l \le d$. The elastic modulus $\mathbb{A}$ is isotropic, constant, and positive definite (notice that $\boldsymbol{e} : \mathbb{A} : \boldsymbol{e} = 2\mu\, \boldsymbol{e} : \boldsymbol{e} + \lambda\, \mathrm{tr}(\boldsymbol{e})^2$ for all $\boldsymbol{e} \in \mathbb{R}^{d\times d}_{\mathrm{sym}}$). □

Owing to the second principle of thermodynamics, the (logarithmic) stress tensor $\boldsymbol{T} \in \mathbb{R}^{d\times d}_{\mathrm{sym}}$ and the internal forces $\boldsymbol{q} \in \mathbb{R}^m$ are derived from Ψ as follows:

$$\boldsymbol{T}(\boldsymbol{e}) := \partial_{\boldsymbol{e}}\Psi(\boldsymbol{e}) = \mathbb{A} : \boldsymbol{e}, \qquad \boldsymbol{q}(\boldsymbol{a}) := \partial_{\boldsymbol{a}}\Psi^{\mathrm{p}}(\boldsymbol{a}). \tag{7.5}$$

(Notice that $\boldsymbol{T}(\boldsymbol{\varepsilon}(\boldsymbol{u})) = \mathbb{A} : \boldsymbol{\varepsilon}(\boldsymbol{u})$ coincides with the usual stress tensor in the case of infinitesimal deformations and no plasticity.)

The criterion to determine whether the deformations are plastic hinges on a scalar yield function $\Phi : \mathbb{R}^{d\times d}_{\mathrm{sym}} \times \mathbb{R}^m \to \mathbb{R}$, which is a continuous and convex function of the stress tensor $\boldsymbol{T}$ and the internal forces $\boldsymbol{q}$. The convex set of admissible states (or plasticity admissible domain) is

$$\mathcal{A} := \left\{(\boldsymbol{T}, \boldsymbol{q}) \in \mathbb{R}^{d\times d}_{\mathrm{sym}} \times \mathbb{R}^m \mid \Phi(\boldsymbol{T}, \boldsymbol{q}) \le 0\right\}. \tag{7.6}$$

This set is partitioned into the elastic domain $\mathcal{A}^{\mathrm{e}} := \{(\boldsymbol{T}, \boldsymbol{q}) \in \mathcal{A} \mid \Phi(\boldsymbol{T}, \boldsymbol{q}) < 0\} = \mathrm{int}(\mathcal{A})$ and the yield surface $\partial\mathcal{A} := \{(\boldsymbol{T}, \boldsymbol{q}) \in \mathcal{A} \mid \Phi(\boldsymbol{T}, \boldsymbol{q}) = 0\}$.

Hypothesis 7.2 (*Yield function*) The yield function satisfies the following properties: (i) Φ is piecewise analytical; (ii) the point $(\boldsymbol{0}, \boldsymbol{0})$ lies in the elastic domain, i.e., $\Phi(\boldsymbol{0}, \boldsymbol{0}) < 0$; (iii) Φ is differentiable at all points on the yield surface $\partial\mathcal{A}$. □

Example 7.3 (*Nonlinear isotropic hardening with von Mises yield criterion*) The internal variable is $\boldsymbol{\alpha} := p$, where $p \ge 0$ is the equivalent plastic strain. The plastic part of the free energy is $\Psi^{\mathrm{p}}(p) := \sigma_{y,0}p + \frac{H}{2}p^2 + (\sigma_{y,\infty} - \sigma_{y,0})(p - \frac{1-e^{-\delta p}}{\delta})$, where $H \ge 0$ is the isotropic hardening modulus, $\sigma_{y,0} > 0$, resp. $\sigma_{y,\infty} \ge 0$, is the initial, resp. infinite, yield stress and $\delta \ge 0$ is the saturation parameter. The internal force is $q := \sigma_{y,0} + Hp + (\sigma_{y,\infty} - \sigma_{y,0})(1 - e^{-\delta p})$. The perfect plasticity model is retrieved by taking $H := 0$ and $\sigma_{y,\infty} := \sigma_{y,0}$. Finally, the J_2-plasticity model with a von Mises criterion uses the yield function $\Phi(\boldsymbol{T}, q) := \sqrt{3/2}\|\,\mathrm{dev}(\boldsymbol{T})\|_{\ell^2} - q$, where $\mathrm{dev}(\boldsymbol{t}) := \boldsymbol{t} - \frac{1}{d}\,\mathrm{tr}(\boldsymbol{t})\boldsymbol{I}_d$, $\|\boldsymbol{t}\|_{\ell^2} := (\boldsymbol{t} : \boldsymbol{t})^{\frac{1}{2}}$ for any tensor $\boldsymbol{t} \in \mathbb{R}^{d\times d}$.

7.1.3 Plasticity Problem in Incremental Form

We are interested in finding the quasi-static evolution in the pseudo-time interval $J := [0, T_\mathrm{f}]$, $T_\mathrm{f} > 0$, of the elastoplastic material body. We focus on the incremental form of the problem so that J is discretized into N subintervals defined by the discrete pseudo-time nodes $t^0 := 0 < t^1 < \cdots < t^N := T_\mathrm{f}$. The evolution occurs, for all $1 \le n \le N$, under the action of a body force $\boldsymbol{f}^n : \Omega \to \mathbb{R}^d$, a traction force $\boldsymbol{g}_\mathrm{N}^n : \partial\Omega_\mathrm{N} \to \mathbb{R}^d$ on the Neumann boundary $\partial\Omega_\mathrm{N}$, and a prescribed displacement $\boldsymbol{u}_\mathrm{D}^n : \partial\Omega_\mathrm{D} \to \mathbb{R}^d$ on the Dirichlet boundary $\partial\Omega_\mathrm{D}$ ($\partial\Omega_\mathrm{D}$ has positive measure to prevent rigid-body motions). Recalling that $\boldsymbol{H}^1(\Omega) := H^1(\Omega; \mathbb{R}^d)$, we denote by $\boldsymbol{V}_\mathrm{D}^n$, resp. $\boldsymbol{V}_0$, the set of all kinematically admissible displacements which satisfy the Dirichlet conditions, resp. homogeneous Dirichlet conditions on $\partial\Omega_\mathrm{D}$:

$$\boldsymbol{V}_\mathrm{D}^n = \left\{\boldsymbol{v} \in \boldsymbol{H}^1(\Omega) \mid \boldsymbol{v}_{|\partial\Omega_\mathrm{D}} = \boldsymbol{u}_\mathrm{D}^n\right\}, \quad \boldsymbol{V}_0 = \left\{\boldsymbol{v} \in \boldsymbol{H}^1(\Omega) \mid \boldsymbol{v}_{|\partial\Omega_\mathrm{D}} = \boldsymbol{0}\right\}. \tag{7.7}$$

It is customary to regroup the plastic strain tensor and the internal variables into the so-called generalized internal variables so that

$$\boldsymbol{\chi} := (\boldsymbol{E}^\mathrm{p}, \boldsymbol{\alpha}) \in \mathcal{X} := \left\{(\boldsymbol{p}, \boldsymbol{\alpha}) \in \mathbb{R}^{d\times d}_\mathrm{sym} \times \mathbb{R}^m \mid \mathrm{tr}(\boldsymbol{p}) = 0\right\}. \tag{7.8}$$

The incremental plasticity problem proceeds as follows: For all $1 \le n \le N$, given $\boldsymbol{u}^{n-1} \in \boldsymbol{V}_\mathrm{D}^{n-1}$ and $\boldsymbol{\chi}^{n-1} := (\boldsymbol{E}^{\mathrm{p},n-1}, \boldsymbol{\alpha}^{n-1}) \in L^2(\Omega; \mathcal{X})$ from the previous pseudo-time step or the initial condition, find $\boldsymbol{u}^n \in \boldsymbol{V}_\mathrm{D}^n$ and $\boldsymbol{\chi}^n := (\boldsymbol{E}^{\mathrm{p},n}, \boldsymbol{\alpha}^n) \in L^2(\Omega; \mathcal{X})$ such that

$$(\boldsymbol{P}^n, \nabla \boldsymbol{w})_{L^2(\Omega)} = \ell^n(\boldsymbol{w}) := (\boldsymbol{f}^n, \boldsymbol{w})_{L^2(\Omega)} + (\boldsymbol{g}_\mathrm{N}^n, \boldsymbol{w})_{L^2(\partial\Omega_\mathrm{N})}, \quad \forall \boldsymbol{w} \in \boldsymbol{V}_0, \tag{7.9}$$

$$(\boldsymbol{\chi}^n, \boldsymbol{P}^n) := \texttt{PLASTICITY}(\boldsymbol{\chi}^{n-1}, \boldsymbol{F}^{n-1}, \boldsymbol{F}^n) \quad \text{pointwise in } \Omega, \tag{7.10}$$

where $\boldsymbol{F}^m := \boldsymbol{F}(\boldsymbol{u}^m)$, $m \in \{n-1, n\}$. Letting $\boldsymbol{E}^m := \mathcal{L}(\boldsymbol{F}^m)$, the procedure `PLASTICITY` finds $\boldsymbol{\chi}^n$ and the Lagrange multiplier Λ^n solving the following constrained nonlinear problem:

$$\boldsymbol{E}^{\mathrm{p},n} - \boldsymbol{E}^{\mathrm{p},n-1} = \Lambda^n \partial_{\boldsymbol{T}} \Phi(\boldsymbol{T}^n, \boldsymbol{q}^n), \quad \boldsymbol{\alpha}^n - \boldsymbol{\alpha}^{n-1} = -\Lambda^n \partial_{\boldsymbol{q}} \Phi(\boldsymbol{T}^n, \boldsymbol{q}^n), \tag{7.11}$$

$$\Lambda^n \ge 0, \quad \Phi(\boldsymbol{T}^n, \boldsymbol{q}^n) \le 0, \quad \Lambda^n \Phi(\boldsymbol{T}^n, \boldsymbol{q}^n) = 0, \tag{7.12}$$

where $\boldsymbol{T}^n := \partial_{\boldsymbol{e}} \Psi(\boldsymbol{E}^n - \boldsymbol{E}^{\mathrm{p},n}) = \mathbb{A} : (\boldsymbol{E}^n - \boldsymbol{E}^{\mathrm{p},n})$ and $\boldsymbol{q}^n := \partial_{\boldsymbol{a}} \Psi^\mathrm{p}(\boldsymbol{\alpha}^n)$. The first Piola–Kirchhoff stress tensor is then defined as $\boldsymbol{P}^n := \boldsymbol{T}^n : \partial_t \mathcal{L}(\boldsymbol{E}^n)$, noting that for infinitesimal deformations, $\boldsymbol{P}^n \approx \mathbb{A} : (\boldsymbol{E}^n - \boldsymbol{E}^{\mathrm{p},n})$. One example of procedure for solving (7.11)–(7.12) is the standard radial return mapping [134, 135]. For strain-hardening plasticity and infinitesimal deformations, the weak formulation (7.9)-(7.10) is well-posed, see [103, Sect. 6.4]. For perfect plasticity, under additional hypotheses on the loads, the existence of a solution with bounded infinitesimal deformation is studied in [70].

The incremental problem (7.9)–(7.10) can be reformulated as an incremental variational inequality by introducing a dissipative function [84, 123]. Given $(\boldsymbol{u}^{n-1}, \boldsymbol{\chi}^{n-1}) \in \boldsymbol{V}_{\mathrm{D}}^{n-1} \times L^2(\Omega; \mathcal{X})$, we define the energy functional $\mathfrak{E}^n : \boldsymbol{V}_{\mathrm{D}}^{n} \times L^2(\Omega; \mathcal{X}) \to \mathbb{R}$ such that

$$\mathfrak{E}^n(\boldsymbol{v}, \boldsymbol{\theta}) := \int_\Omega \Psi^n(\boldsymbol{F}(\boldsymbol{v}), \boldsymbol{\theta})\,\mathrm{d}\boldsymbol{x} - \ell^n(\boldsymbol{v}), \tag{7.13}$$

with the incremental pseudo-energy density $\Psi^n : \mathbb{R}_+^{d\times d} \times \mathcal{X} \to \mathbb{R}$ such that

$$\Psi^n(\boldsymbol{F}, \boldsymbol{\theta}) := \Psi(\boldsymbol{E}^{\mathrm{e}}, \boldsymbol{\alpha}) - \Psi(\boldsymbol{E}^{\mathrm{e},n-1}, \boldsymbol{\alpha}^{n-1}) + D_{\boldsymbol{\chi}^{n-1}}(\boldsymbol{\theta}), \tag{7.14}$$

where $\boldsymbol{\theta} := (\boldsymbol{E}^{\mathrm{p}}, \boldsymbol{\alpha})$, $\boldsymbol{E}^{\mathrm{e}} := \boldsymbol{E} - \boldsymbol{E}^{\mathrm{p}}$ with $\boldsymbol{E} := \mathcal{L}(\boldsymbol{F})$, and with the incremental dissipation function $D_{\boldsymbol{\chi}^{n-1}}(\boldsymbol{\theta}) := \sup_{(\boldsymbol{T},\boldsymbol{q})\in\mathcal{A}} \big((\boldsymbol{T} : (\boldsymbol{E}^{\mathrm{p}} - \boldsymbol{E}^{\mathrm{p},n-1}) - \boldsymbol{q}\cdot(\boldsymbol{\alpha} - \boldsymbol{\alpha}^{n-1})\big)$ (D is convex and positively homogeneous of degree one). Then, a pair $(\boldsymbol{u}^n, \boldsymbol{\chi}^n) \in \boldsymbol{V}_{\mathrm{D}}^{n} \times L^2(\Omega; \mathcal{X})$ solving (7.9)–(7.10) satisfies the Euler–Lagrange equations of the minimization problem $\min_{(\boldsymbol{v},\boldsymbol{\theta})\in \boldsymbol{V}_{\mathrm{D}}^{n}\times L^2(\Omega;\mathcal{X})} \mathfrak{E}^n(\boldsymbol{v}, \boldsymbol{\theta})$.

7.2 HHO Discretizations

In this section, we present HHO methods to solve nonlinear plasticity problems.

7.2.1 Discrete Unknowns

Let $\mathcal{T}$ be a mesh of Ω belonging to a shape-regular mesh sequence (see Sects. 1.2.1 and 2.1.1). We assume that Ω is a polyhedron so that the mesh covers Ω exactly. Moreover, we assume that every mesh boundary face belongs either to $\partial\Omega_{\mathrm{D}}$ or to $\partial\Omega_{\mathrm{N}}$. The corresponding subsets of $\mathcal{F}^\partial$ are denoted by $\mathcal{F}_{\mathrm{D}}^\partial$ and $\mathcal{F}_{\mathrm{N}}^\partial$. Recall that in HHO methods, the discrete unknowns are polynomials attached to the mesh cells and the mesh faces. In the context of continuum mechanics, both unknowns are vector-valued: the cell unknowns approximate the displacement field in the cell, and the face unknowns approximate its trace on the mesh faces; see Fig. 4.1.

For simplicity, we consider only the equal order-case where $k \geq 1$ is the polynomial degree of both face and cell unknowns. For every mesh cell $T \in \mathcal{T}$, we set

$$\hat{\boldsymbol{V}}_T^k := \mathbb{P}_d^k(T) \times \mathbb{P}_{d-1}^k(\mathcal{F}_T), \qquad \mathbb{P}_{d-1}^k(\mathcal{F}_T) := \bigtimes_{F\in\mathcal{F}_T} \mathbb{P}_{d-1}^k(F), \tag{7.15}$$

with $\mathbb{P}_d^k(T) := \mathbb{P}_d^k(T;\mathbb{R}^d)$ and $\mathbb{P}_{d-1}^k(F) := \mathbb{P}_{d-1}^k(F;\mathbb{R}^d)$. A generic element in $\hat{\boldsymbol{V}}_T^k$ is denoted by $\hat{\boldsymbol{v}}_T := (\boldsymbol{v}_T, \boldsymbol{v}_{\partial T})$. The HHO space is then defined as follows:

$$\hat{\boldsymbol{V}}_h^k := \boldsymbol{V}_{\mathcal{T}}^k \times \boldsymbol{V}_{\mathcal{F}}^k, \qquad \boldsymbol{V}_{\mathcal{T}}^k := \bigtimes_{T \in \mathcal{T}} \mathbb{P}_d^k(T), \qquad \boldsymbol{V}_{\mathcal{F}}^k := \bigtimes_{F \in \mathcal{F}} \mathbb{P}_{d-1}^k(F). \tag{7.16}$$

A generic element in $\hat{\boldsymbol{V}}_h^k$ is denoted by $\hat{\boldsymbol{v}}_h := (\boldsymbol{v}_{\mathcal{T}}, \boldsymbol{v}_{\mathcal{F}})$ with $\boldsymbol{v}_{\mathcal{T}} := (\boldsymbol{v}_T)_{T \in \mathcal{T}}$ and $\boldsymbol{v}_{\mathcal{F}} := (\boldsymbol{v}_F)_{F \in \mathcal{F}}$, and we localize the components of $\hat{\boldsymbol{v}}_h$ associated with a mesh cell $T \in \mathcal{T}$ and its faces by using the notation $\hat{\boldsymbol{v}}_T := \big(\boldsymbol{v}_T, \boldsymbol{v}_{\partial T} := (\boldsymbol{v}_F)_{F \in \mathcal{F}_T}\big) \in \hat{\boldsymbol{V}}_T^k$. The Dirichlet boundary condition on the displacement field is enforced explicitly on the discrete unknowns attached to the mesh boundary faces in $\mathcal{F}_{\mathrm{D}}^{\partial}$. Letting $\boldsymbol{\Pi}_F^k$ denote the L^2-orthogonal projection onto $\mathbb{P}_{d-1}^k(F)$, we set

$$\hat{\boldsymbol{V}}_{h,\mathrm{D}}^{k,n} := \left\{ \hat{\boldsymbol{v}}_h \in \hat{\boldsymbol{V}}_h^k \mid \boldsymbol{v}_F = \boldsymbol{\Pi}_F^k(\boldsymbol{u}_{\mathrm{D}}^n),\ \forall F \in \mathcal{F}_{\mathrm{D}}^{\partial} \right\}, \tag{7.17}$$

$$\hat{\boldsymbol{V}}_{h,0}^{k} := \left\{ \hat{\boldsymbol{v}}_h \in \hat{\boldsymbol{V}}_h^k \mid \boldsymbol{v}_F = \boldsymbol{0},\ \forall F \in \mathcal{F}_{\mathrm{D}}^{\partial} \right\}. \tag{7.18}$$

The discrete generalized internal variables are computed locally at the quadrature points of every mesh cell. We introduce the quadrature points $\boldsymbol{\xi}_T = (\boldsymbol{\xi}_{T,j})_{1 \le j \le m_{\mathrm{Q}}}$ and the weights $\boldsymbol{\omega}_T = (\omega_{T,j})_{1 \le j \le m_{\mathrm{Q}}}$, with $\boldsymbol{\xi}_{T,j} \in T$ and $\omega_{T,j} \in \mathbb{R}$ for all $1 \le j \le m_{\mathrm{Q}}$ and all $T \in \mathcal{T}$. We denote by k_{Q} the order of the quadrature. Then, the discrete generalized internal variables are sought in the space

$$\mathcal{X}_{\mathcal{T}} := \bigtimes_{T \in \mathcal{T}} \big(\underbrace{\mathcal{X} \times \cdots \times \mathcal{X}}_{m_{\mathrm{Q}} \text{times}}\big), \tag{7.19}$$

that is, for all $T \in \mathcal{T}$, the generalized internal variables attached to T form a vector $\boldsymbol{\chi}_T$ whose components are (a bit abusively) denoted by $(\boldsymbol{\chi}_T(\boldsymbol{\xi}_{T,j}))_{1 \le j \le m_{\mathrm{Q}}}$ with $\boldsymbol{\chi}_T(\boldsymbol{\xi}_{T,j}) \in \mathcal{X}$ for all $1 \le j \le m_{\mathrm{Q}}$. In what follows, we use the following notation:

$$(\boldsymbol{p}, \boldsymbol{q})_{L^2_{\mathrm{Q}}(T)} := \sum_{j=1}^{m_{\mathrm{Q}}} \omega_{T,j}\, \boldsymbol{p}(\boldsymbol{\xi}_{T,j}) : \boldsymbol{q}(\boldsymbol{\xi}_{T,j}), \tag{7.20}$$

where, according to the context, the arguments can be either a continuous, tensor-valued function defined on T or a vector in $(\mathbb{R}^{d \times d})^{m_{\mathrm{Q}}}$. The global counterpart $(\boldsymbol{p}, \boldsymbol{q})_{L^2_{\mathrm{Q}}(\Omega)}$ is obtained by summing (7.20) over the mesh cells.

7.2.2 Discrete Plasticity Problem in Incremental Form

Recall the local gradient reconstruction $\boldsymbol{G}_T : \hat{\boldsymbol{V}}_T^k \to \mathbb{P}^k(T; \mathbb{R}^{d \times d})$ defined in (4.52) and the deformation gradient operator such that $\boldsymbol{F}_T(\hat{\boldsymbol{v}}_T) := \boldsymbol{I}_d + \boldsymbol{G}_T(\hat{\boldsymbol{v}}_T)$ for all $T \in \mathcal{T}$. The global counterparts of these operators, which are defined in every mesh cell as above, are tensor-valued piecewise polynomials in $\mathbb{P}^k(\mathcal{T}; \mathbb{R}^{d \times d})$ denoted by $\boldsymbol{G}_{\mathcal{T}}$ and $\boldsymbol{F}_{\mathcal{T}}$. The global stabilization bilinear form $s_h : \hat{\boldsymbol{V}}_h^k \times \hat{\boldsymbol{V}}_h^k \to \mathbb{R}$ is defined in

(4.36) as for the linear elasticity problem, and we consider a positive weight $\beta_0 > 0$ (the choice $\beta_0 = 1$ was made for linear elasticity in Sect. 4.2.2).

The discrete plasticity problem in incremental form proceeds as follows: For all $1 \le n \le N$, given $\hat{\boldsymbol{u}}_h^{n-1} \in \hat{\boldsymbol{V}}_{h,\mathrm{D}}^{k,n-1}$ and $\boldsymbol{\chi}_{\mathcal{T}}^{n-1} := (\boldsymbol{E}_{\mathcal{T}}^{\mathrm{p},n-1}, \boldsymbol{\alpha}_{\mathcal{T}}^{n-1}) \in \mathcal{X}_{\mathcal{T}}$ from the previous pseudo-time step or the initial condition, find $\hat{\boldsymbol{u}}_h^n \in \hat{\boldsymbol{V}}_{h,\mathrm{D}}^{k,n}$ and $\boldsymbol{\chi}_{\mathcal{T}}^n := (\boldsymbol{E}_{\mathcal{T}}^{\mathrm{p},n}, \boldsymbol{\alpha}_{\mathcal{T}}^n) \in \mathcal{X}_{\mathcal{T}}$ such that

$$(\boldsymbol{P}_{\mathcal{T}}^n, \boldsymbol{G}_{\mathcal{T}}(\hat{\boldsymbol{w}}_h))_{L^2_{\mathrm{Q}}(\Omega)} + 2\beta_0\mu s_h(\hat{\boldsymbol{u}}_h^n, \hat{\boldsymbol{w}}_h) = \ell^n(\hat{\boldsymbol{w}}_h), \quad \forall \hat{\boldsymbol{w}}_h \in \hat{\boldsymbol{V}}_{h,0}^k, \tag{7.21}$$

$$(\boldsymbol{\chi}_{T,j}^n, \boldsymbol{P}_{T,j}^n) := \texttt{PLASTICITY}(\boldsymbol{\chi}_{T,j}^{n-1}, \boldsymbol{F}_{T,j}^{n-1}, \boldsymbol{F}_{T,j}^n), \ \forall T \in \mathcal{T}, \forall j \in \{1, \dots, m_{\mathrm{Q}}\}, \tag{7.22}$$

where $\boldsymbol{\chi}_{T,j}^m := \boldsymbol{\chi}_T^m(\boldsymbol{\xi}_{T,j})$, $\boldsymbol{F}_{T,j}^m := \boldsymbol{F}_{\mathcal{T}}^m(\boldsymbol{\xi}_{T,j})$ with $m \in \{n-1, n\}$, and $\boldsymbol{P}_{\mathcal{T}}^n(\boldsymbol{\xi}_{T,j}) := \boldsymbol{P}_{T,j}^n$ for all $T \in \mathcal{T}$ and all $j \in \{1, \dots, m_{\mathrm{Q}}\}$. Notice that the same procedure `PLASTICITY` is used as in the continuous setting.

Remark 7.4 (*Literature*) HHO methods for plasticity were developed in [2, 3]. Discontinuous Galerkin methods have been developed in [104, 120, 121], and virtual element methods in [9, 15, 150].

7.2.3 Nonlinear Solver

The nonlinear problem (7.21)–(7.22) can be solved by using Newton's method. This requires evaluating the consistent (nominal) elastoplastic tangent modulus $\mathbb{A}_{\mathrm{ep}}$ at every Gauss point in every mesh cell. The evaluation of $\mathbb{A}_{\mathrm{ep}}$ can be included within the procedure `PLASTICITY`. To this purpose, we rewrite (7.10) as

$$(\boldsymbol{\chi}^n, \boldsymbol{P}^n, \mathbb{A}_{\mathrm{ep}}^n) := \texttt{PLASTICITY}(\boldsymbol{\chi}^{n-1}, \boldsymbol{F}^{n-1}, \boldsymbol{F}^n). \tag{7.23}$$

Referring to the constrained nonlinear problem (7.11)–(7.12) and recalling that $\mathbb{A}$ denotes the (state-independent) elastic modulus (see (7.1)), one first computes the infinitesimal elastoplastic tangent modulus $\mathbb{A}_{\mathrm{ep},\infty}^n$ such that

$$\mathbb{A}_{\mathrm{ep},\infty}^n := \mathbb{A} - \frac{(\mathbb{A} : \partial_{\boldsymbol{T}}\Phi) \otimes (\mathbb{A} : \partial_{\boldsymbol{T}}\Phi)}{\partial_{\boldsymbol{T}}\Phi : \mathbb{A} : \partial_{\boldsymbol{T}}\Phi + \partial_{\boldsymbol{q}}\Phi : \partial_{\boldsymbol{aa}}^2\Psi^{\mathrm{p}} : \partial_{\boldsymbol{q}}\Phi}, \tag{7.24}$$

with the partial derivatives of Φ evaluated at $(\boldsymbol{T}^n, \boldsymbol{q}^n)$ and the second derivative of Ψ^{p} evaluated at $\boldsymbol{\alpha}^n$. Then, one sets

$$\mathbb{A}_{\mathrm{ep}}^n := (\partial_t\mathcal{L})^{\mathsf{T}} : \mathbb{A}_{\mathrm{ep},\infty}^n : \partial_t\mathcal{L} + \boldsymbol{T}^n : \partial_{tt}^2\mathcal{L}, \tag{7.25}$$

where the partial derivatives of $\mathcal{L}$ are evaluated at $\boldsymbol{E}^n$.

Let $i \geq 0$ be the index of the Newton's iteration and recall that $\hat{\boldsymbol{u}}_h^{n-1} \in \hat{\boldsymbol{V}}_{h,\mathrm{D}}^{k,n-1}$ and $\boldsymbol{\chi}_{\mathcal{T}}^{n-1} \in \mathcal{X}_{\mathcal{T}}$ are given from the previous pseudo-time step or the initial condition. The Newton's method is initialized by setting $\hat{\boldsymbol{u}}_h^{n,0} := \hat{\boldsymbol{u}}_h^{n-1}$ (up to the update of the Dirichlet condition) and $\boldsymbol{\chi}_{\mathcal{T}}^{n,0} := \boldsymbol{\chi}_{\mathcal{T}}^{n-1}$. Then, for all $i \geq 0$, given $\hat{\boldsymbol{u}}_h^{n,i} \in \hat{\boldsymbol{V}}_{h,\mathrm{D}}^{k,n}$, one computes at each Newton's iteration the incremental displacement $\delta\hat{\boldsymbol{u}}_h^{n,i} \in \hat{\boldsymbol{V}}_{h,0}^{k}$ such that

$$(\mathbb{A}_{\mathrm{ep},\mathcal{T}}^{n,i}:\boldsymbol{G}_{\mathcal{T}}(\delta\hat{\boldsymbol{u}}_h^{n,i}), \boldsymbol{G}_{\mathcal{T}}(\hat{\boldsymbol{w}}_h))_{\boldsymbol{L}_{\mathrm{Q}}^2(\Omega)} + 2\beta_0\mu s_h(\delta\hat{\boldsymbol{u}}_h^{n,i}, \hat{\boldsymbol{w}}_h) = -R_h^{n,i}(\hat{\boldsymbol{w}}_h), \tag{7.26}$$

$$(\boldsymbol{\chi}_{T,j}^{n,i}, \boldsymbol{P}_{T,j}^{n,i}, \mathbb{A}_{\mathrm{ep},T,j}^{n,i}) := \texttt{PLASTICITY}(\boldsymbol{\chi}_{T,j}^{n-1}, \boldsymbol{F}_{T,j}^{n-1}, \boldsymbol{F}_{T,j}^{n,i}), \tag{7.27}$$

where (7.26) holds for all $\hat{\boldsymbol{w}}_h \in \hat{\boldsymbol{V}}_{h,0}^{k}$ with the residual term

$$R_h^{n,i}(\hat{\boldsymbol{w}}_h) := (\boldsymbol{P}_{\mathcal{T}}^{n,i}, \boldsymbol{G}_{\mathcal{T}}(\hat{\boldsymbol{w}}_h))_{\boldsymbol{L}_{\mathrm{Q}}^2(\Omega)} + 2\beta_0\mu s_h(\hat{\boldsymbol{u}}_h^{n,i}, \hat{\boldsymbol{w}}_h) - \ell^n(\hat{\boldsymbol{w}}_h), \tag{7.28}$$

and where (7.27) holds for all $T \in \mathcal{T}$ and all $j \in \{1, \dots, m_{\mathrm{Q}}\}$, with $\boldsymbol{F}_{T,j}^{n-1}$, $\boldsymbol{F}_{T,j}^{n,i}$ evaluated from $\hat{\boldsymbol{u}}_h^{n-1}$, $\hat{\boldsymbol{u}}_h^{n,i}$, respectively, and $\mathbb{A}_{\mathrm{ep},\mathcal{T}}^{n,i}(\boldsymbol{\xi}_{T,j}) := \mathbb{A}_{\mathrm{ep},T,j}^{n,i}$. At the end of each Newton's iteration, one updates the discrete displacement as $\hat{\boldsymbol{u}}_h^{n,i+1} = \hat{\boldsymbol{u}}_h^{n,i} + \delta\hat{\boldsymbol{u}}_h^{n,i}$. The discrete generalized internal variables do not need to be updated at the end of the iteration, but only once Newton's method has converged.

For strain-hardening plasticity, the consistent elastoplastic tangent modulus is symmetric positive-definite. The following result gives some sufficient conditions for the linear system (7.26) to be coercive.

Theorem 7.5 (Coercivity) *Assume the following:* (i) $k_{\mathrm{Q}} \geq 2k$ *and all the quadrature weights are positive;* (ii) $\beta_0 > 0$; (iii) *the plastic model is strain-hardening. Let* $\theta > 0$ *be the smallest eigenvalue of the fourth-order symmetric positive-definite tensors* $(2\mu)^{-1}\mathbb{A}_{ep,T,j}^{n,i}$ *for all* $T \in \mathcal{T}$ *and all* $j \in \{1, \dots, m_{\mathrm{Q}}\}$. *Then, the linear system* (7.26) *in each Newton's iteration is coercive, i.e., there is* $\alpha > 0$, *independent of* h, *such that for all* $\hat{\boldsymbol{v}}_h \in \hat{\boldsymbol{V}}_{h,0}^{k}$,

$$(\mathbb{A}_{ep,\mathcal{T}}^{n,i}:\boldsymbol{G}_{\mathcal{T}}(\hat{\boldsymbol{v}}_h), \boldsymbol{G}_{\mathcal{T}}(\hat{\boldsymbol{v}}_h))_{\boldsymbol{L}_{\mathrm{Q}}^2(\Omega)} + 2\beta_0\mu s_h(\hat{\boldsymbol{v}}_h, \hat{\boldsymbol{v}}_h) \geq \alpha \min(\beta_0, \theta) 2\mu \|\hat{\boldsymbol{v}}_h\|_{\hat{\boldsymbol{V}}_{h,0}^k}^2, \tag{7.29}$$

where $\|\hat{\boldsymbol{v}}_h\|_{\hat{\boldsymbol{V}}_{h,0}^k}^2 := \sum_{T\in\mathcal{T}} |\hat{\boldsymbol{v}}_T|_{\hat{\boldsymbol{V}}_T^k}^2$, $|\hat{\boldsymbol{v}}_T|_{\hat{\boldsymbol{V}}_T^k}^2 := \|\nabla \boldsymbol{v}_T\|_{\boldsymbol{L}^2(T)}^2 + h_T^{-1}\|\boldsymbol{v}_T - \boldsymbol{v}_{\partial T}\|_{\boldsymbol{L}^2(\partial T)}^2$.

Proof Since the material is strain-hardening, we have $\theta > 0$. Let $\hat{\boldsymbol{v}}_h \in \hat{\boldsymbol{V}}_{h,0}^{k}$. Since $\boldsymbol{G}_T(\hat{\boldsymbol{v}}_T) \in \mathbb{P}_d^k(T, \mathbb{R}^{d\times d})$ for all $T \in \mathcal{T}$, since all the quadrature weights are positive, and $k_{\mathrm{Q}} \geq 2k$, we infer that

$$
\begin{aligned}
&(\mathbb{A}^{n,i}_{\mathrm{ep},\mathcal{T}}:\boldsymbol{G}_{\mathcal{T}}(\hat{\boldsymbol{v}}_h), \boldsymbol{G}_{\mathcal{T}}(\hat{\boldsymbol{v}}_h))_{\boldsymbol{L}^2_{\mathrm{Q}}(\Omega)} + 2\beta_0\mu s_h(\hat{\boldsymbol{v}}_h, \hat{\boldsymbol{v}}_h) \\
&\geq \sum_{T\in\mathcal{T}} \sum_{1\leq j\leq m_{\mathrm{Q}}} 2\mu\theta\omega_{T,j} \|\boldsymbol{G}_T(\hat{\boldsymbol{v}}_T)(\boldsymbol{\xi}_{T,j})\|^2_{\ell^2} + 2\beta_0\mu s_h(\hat{\boldsymbol{v}}_h, \hat{\boldsymbol{v}}_h) \\
&\geq 2\mu \min(\theta, \beta_0) \sum_{T\in\mathcal{T}} \Big(\|\boldsymbol{G}_T(\hat{\boldsymbol{v}}_T)\|^2_{\boldsymbol{L}^2(T)} + h_T^{-1}\|\boldsymbol{S}_{\partial T}(\hat{\boldsymbol{v}}_T)\|^2_{\boldsymbol{L}^2(\partial T)} \Big).
\end{aligned}
$$

We conclude by using the stability result from Lemma 4.11.

Remark 7.6 (*Choice of* β_0) Theorem 7.5 indicates that the smallest eigenvalue θ is a natural target for the value of the weight parameter β_0 in the stabilization. A numerical study on the influence of β_0 is presented in [3, Sect. 5.3]. Another possibility considered for virtual element methods in [150] is a piecewise constant stabilization parameter depending on the shape of the cell, the value of θ, and a minimal user-defined value when $\theta \leq 0$.

7.3 Numerical Examples

The goal of this section is to illustrate the above HHO method on two industrial applications where finite plasticity is present: a torsion of a square-section bar and an hydraulic pump under internal forces. For both examples, we use the nonlinear isotropic hardening model described in Example 7.3.

7.3.1 Torsion of a Square-Section Bar

This first example allows one to test the robustness of HHO methods under large torsion. The bar has a square-section of length $L := 1$ mm and of height $H := 5$ mm along the z-direction. The bottom end is clamped and the top end is subjected to a rotation of angle $\Theta = 360°$ around its center along the z-direction. The following material parameters ared used: Young modulus $E := 206.9$ GPa, Poisson ratio $\nu := 0.29$, hardening parameter $H := 129.2$ MPa, initial yield stress $\sigma_{y,0} := 450$ MPa, infinite yield stress $\sigma_{y,\infty} := 715$ MPa, and saturation parameter $\delta := 16.93$. The equivalent plastic strain p is plotted at the quadrature points in Fig. 7.1 for $k = 2$. There is no sign of localization of the plastic deformations even for large rotations and large plastic deformations (around 50%). Moreover, the trace of the Cauchy stress tensor $\boldsymbol{\sigma}$ is plotted at the quadrature points on the final configuration in Fig. 7.2 for $k = 2$. As expected, there is no sign of volume locking (no oscillation of the trace of the stress tensor, except at both ends which are fully constrained by Dirichlet conditions, so that stress concentrations are present).

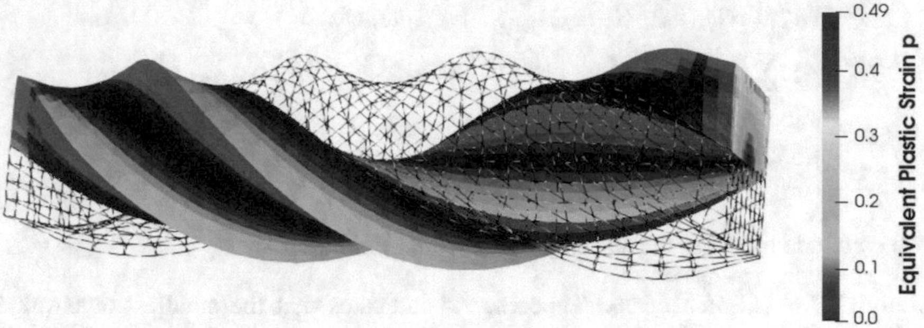

Fig. 7.1 Torsion of a square-section bar: equivalent plastic strain p at the quadrature points for $k = 2$ and a rotation of angle $\Theta = 360°$

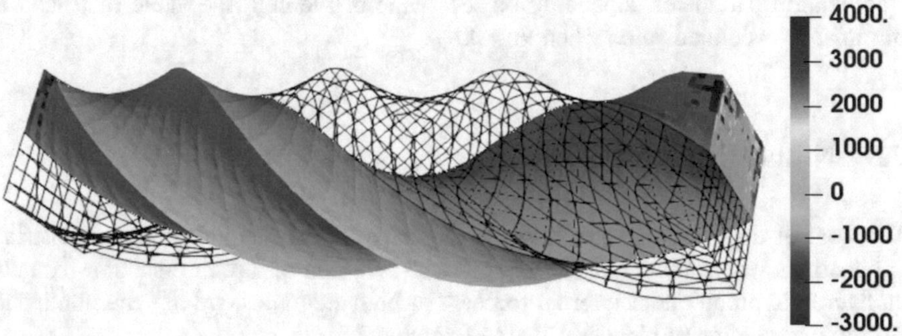

Fig. 7.2 Torsion of a square-section bar: trace of the Cauchy stress tensor $\boldsymbol{\sigma}$ (in MPa) at the quadrature points for $k = 2$ and a rotation of angle $\Theta = 360°$

7.3.2 Hydraulic Pump Under Internal Forces

This test case based on an industrial problem focuses on the deformation of an hydraulic pump and two of its pipes under the influence of a pressurized fluid. Since the study is restricted to the structural part of the problem, the force applied by the fluid on the walls of the pump and its pipes is replaced by an equivalent internal force. This surface force corresponds to a pressure of 14 MPa in the reference configuration. Moreover, the bottom of the pump is clamped and the other surfaces are free. The description of the geometry and the mesh is given on the `code_aster` web site.[1] Strain-hardening plasticity with a von Mises yield criterion is considered with the following material parameters: Young modulus $E := 200$ GPa, Poisson ratio $\nu := 0.3$, hardening parameter $H := 200$ MPa, initial and infinite yield stresses $\sigma_{y,0} = \sigma_{y,\infty} := 500$ MPa, and saturation parameter $\delta := 0$. The mesh is composed of 23,837 tetrahedra and 41,218 triangular faces. The discrete global problem to solve has around 50,000 dofs for $k = 1$. The Euclidean norm of the displacement and the

[1] Test PERF009: https://www.code-aster.org/V2/doc/default/fr/man_v/v1/v1.01.262.pdf.

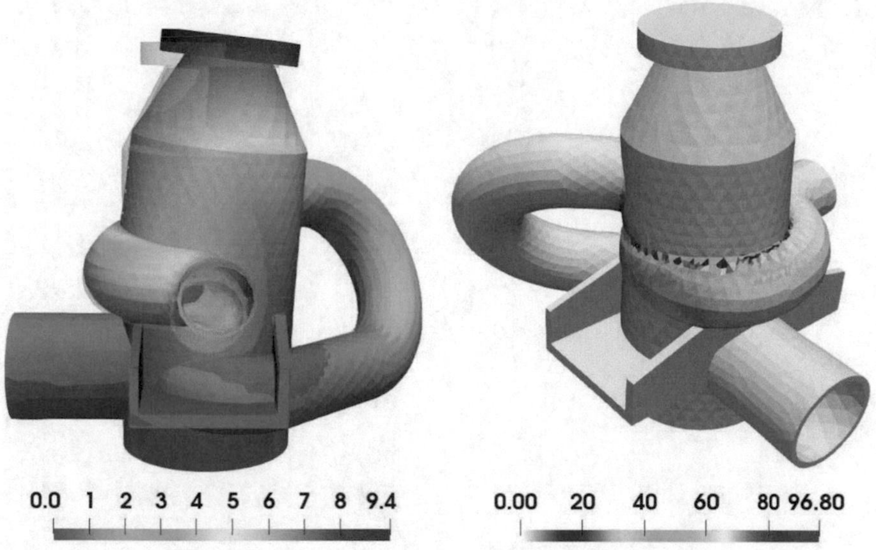

(a) Euclidean norm of the displacement (b) Equivalent plastic strain p

Fig. 7.3 Pump under internal forces. Left: Euclidean norm of the displacement (in mm) for $k = 1$ on the deformed configuration with transparent reference configuration. Right: Equivalent plastic strain p (in %) on the deformed configuration

equivalent plastic strain p are plotted in Fig. 7.3 on the deformed configuration. Note that the upper left part of the pump has the largest displacement. Moreover, we remark that the plastic deformations are mainly present in the pipes and, in particular, at the junction between the pump and its pipes with nearly 97% of equivalent plastic strain p.

Chapter 8
Implementation Aspects

In this chapter, we outline the steps needed to bring the abstract formulation of the HHO method to an actual implementation. For simplicity, we focus on the Poisson model problem (see Chap. 1). We show how the local HHO operators (reconstruction and stabilization) are translated into matrices that can be used in the actual computation, and we give some criteria to test the implementation. Then we discuss the assembly of the discrete problem and the handling of the boundary conditions. We conclude with a brief overview on computational costs. Along the chapter, we provide some snippets of Matlab®/Octave code to show a possible implementation (in 1D) of the critical parts.[1] A 3D/polyhedral code called `DiSk++` fully supporting HHO and discontinuous Galerkin methods is downloadable at the address https://github.com/wareHHOuse/diskpp.[2] We also refer the reader to [58] for a description of the implementation of HHO methods using generic programming.

8.1 Polynomial Spaces

The HHO method employs polynomials attached to the mesh cells and to the mesh faces. These polynomials are represented by their components in chosen polynomial bases. The evaluation of the cell basis functions can be done directly in the physical element by manipulating d-variate polynomials where $d \geq 1$ is the space dimension. Instead, the evaluation of the face basis functions is done by means of affine geometric mappings that transform the d-dimensional points composing a face to a $(d-1)$-dimensional reference system associated with the face so that one manipulates $(d-1)$-variate polynomials; see (1.7).

---

[1] The full source is available at https://github.com/wareHHOuse/demoHHO.

[2] HHO methods are also implemented in the industrial software `code_aster` [90] and the academic codes `SpaFEDTe` and `HArD::Core` available on `github`.

M. Cicuttin et al., *Hybrid High-Order Methods*,
SpringerBriefs in Mathematics,
https://doi.org/10.1007/978-3-030-81477-9_8

Let us consider first the cell basis functions. Let $k \geq 0$ be the polynomial degree and recall that $\mathbb{P}_d^k$ is composed of the d-variate polynomials of total degree at most k with $\dim(\mathbb{P}_d^k) = \binom{k+d}{d} =: N_d^k$. Let $T \in \mathcal{T}$ be a mesh cell and let $\{\phi_{T,i}\}_{1\leq i \leq N_d^k}$ be a basis of $\mathbb{P}_d^k(T)$. Then, any polynomial $p \in \mathbb{P}_d^k(T)$ can be decomposed in this basis as

$$p(\boldsymbol{x}) = \sum_{1\leq i \leq N_d^k} p_i \phi_{T,i}(\boldsymbol{x}), \tag{8.1}$$

where the coefficients $p_i \in \mathbb{R}$ are the components of p in the chosen basis. These coefficients are the actual information that is stored and manipulated during the computations. A simple and useful example of basis functions are the scaled monomials. Let $\boldsymbol{x}_T = (x_{T,i})_{1\leq i\leq d} \in \mathbb{R}^d$ denote the barycenter of T and h_T its diameter. Recall that for a multi-index $\alpha \in \mathbb{N}^d$, $|\alpha| := \sum_{1\leq i\leq d} \alpha_i$ denotes its length. Then, for all $\alpha \in \mathbb{N}^d$ with $|\alpha| \leq k$, we set

$$\mu_{T,\alpha}(\mathbf{x}) := \prod_{1\leq i\leq d} \left(\frac{2(x_i - x_{T,i})}{h_T}\right)^{\alpha_i}, \tag{8.2}$$

leading to the basis $\{\mu_{T,\alpha}\}_{\alpha\in\mathbb{N}^d, |\alpha|\leq k}$ of $\mathbb{P}_d^k(T)$. The two-dimensional scaled monomial basis is depicted in Fig. 8.1 (up to degree 2 and with $h_T = 2$ rather than $2\sqrt{2}$). The code in Listing 8.1 implements (8.2): the function evaluates the basis up to degree max_k and its derivatives in the element with center x_bar and size h. It returns two vectors containing the values of the basis functions and their derivatives at the point x.

Listing 8.1 Possible implementation of a function evaluating the scaled monomial scalar basis and its derivatives in a 1D cell.

```
% Evaluate scalar monomial basis
function [phi, dphi] = basis(x, x_bar, h, max_k)
    k       = (0:max_k)';
    x_tilde = 2*(x-x_bar)/h;

    phi            = x_tilde .^ k;
    dphi           = zeros(max_k+1,1);
    dphi(2:end)    = (2*k(2:end)/h).*(x_tilde.^k(1:end-1));
end
```

The face basis functions can be constructed in an analogous way by working on $\mathbb{R}^{d-1}$ if $d \geq 2$ and using the affine geometric mapping $\boldsymbol{T}_F : \mathbb{R}^{d-1} \to H_F$, where H_F is the affine hyperplane in $\mathbb{R}^d$ supporting F. In particular, scaled monomials can be built by using the point $\boldsymbol{T}_F^{-1}(\boldsymbol{x}_F)$, where $\boldsymbol{x}_F$ is the barycenter of F.

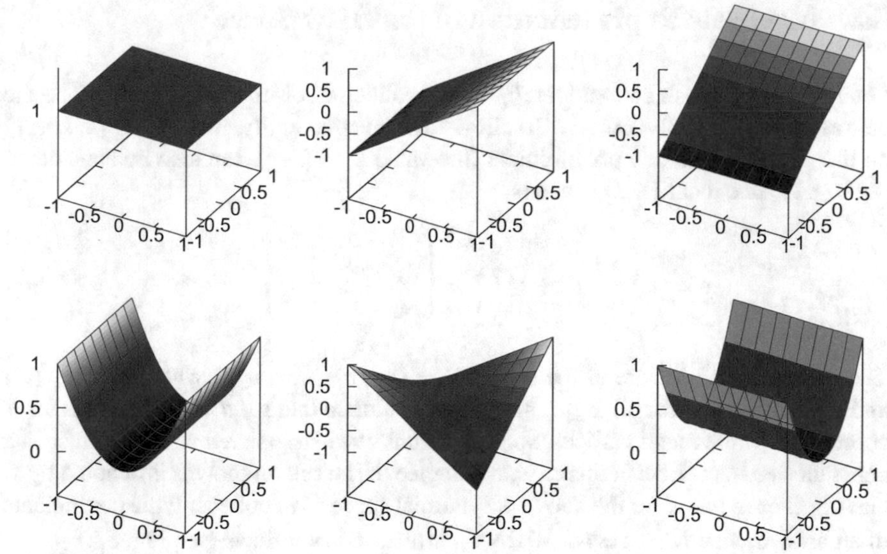

Fig. 8.1 The set of functions of the scaled monomial basis of order 2 in the 2D element $(-1, 1)^2$

Remark 8.1 (*High order*) The choice of the basis functions is particularly important when working with high-order polynomials, and its effects can be seen typically for $k \geq 3$ and beyond (see, e.g., [107, Sect. 3.1] and [92, Sect. 6.3.5 and Remark 7.14] for general discussions). It can be beneficial to work with L^2-orthogonal bases. Such bases are easily devised for $d = 1$ using Legendre polynomials, and for $d \geq 2$ if the cells are rectangular cuboids. If other shapes are used, an orthogonalization procedure can be considered, although it can be expensive. One should bear in mind that the scaled monomial basis suffers from ill-conditioning for high polynomial degrees. □

Remark 8.2 (*Vector-valued case*) In continuum mechanics, HHO methods hinge on vector- and tensor-valued polynomials. Bases for such polynomial spaces can be readily defined as tensor-products of a scalar polynomial basis and the Cartesian basis of $\mathbb{R}^d$ or $\mathbb{R}^{d\times d}$. For example, if we apply this procedure to $\mathbb{P}_2^1$ with the basis $\{1, x, y\}$, we obtain the following vector-valued basis:

$$\begin{bmatrix}1\\0\end{bmatrix}, \begin{bmatrix}0\\1\end{bmatrix}, \begin{bmatrix}x\\0\end{bmatrix}, \begin{bmatrix}0\\x\end{bmatrix}, \begin{bmatrix}y\\0\end{bmatrix}, \begin{bmatrix}0\\y\end{bmatrix}.$$

The same procedure can be readily extended to tensor-valued polynomials. □

8.2 Algebraic Representation of the HHO Space

Let $T \in \mathcal{T}$ be a mesh cell and let $\mathcal{F}_T$ be the collection of its faces. Let $k \geq 0$ be the degree of the face polynomials. To allow for some generality, we let $k' \in \{k, k+1\}$ be the degree of the cell polynomials (the value $k' = k - 1$ can also be considered for $k \geq 1$). The local HHO space is

$$\hat{V}_T^{k',k} := \mathbb{P}_d^{k'}(T) \times \left\{ \bigtimes_{F \in \mathcal{F}_T} \mathbb{P}_{d-1}^k(F) \right\}. \tag{8.3}$$

The members of $\hat{V}_T^{k',k}$ are of the form $\hat{v}_T := (v_T, v_{F_1}, \ldots, v_{F_n})$, where $v_T \in \mathbb{P}_d^{k'}(T)$ and $v_{F_i} \in \mathbb{P}_{d-1}^k(F_i)$ for all $1 \leq i \leq n := \#\mathcal{F}_T$. Notice that for $d = 1$, the mesh faces coincide with the mesh vertices, so that the unknown associated with each face is a constant (see Sect. 1.6); in this case, the degree of the cell unknowns is denoted by k. Having chosen bases for the above polynomial spaces, we collect all the coefficients in an array of size $\hat{N}_d^{k',k} := N_d^{k'} + nN_{d-1}^k$ structured as follows (see Fig. 8.2):

$$\mathsf{v}_T := \left[\mathsf{v}_{T,1}, \ldots, \mathsf{v}_{T,N_d^{k'}} | \mathsf{v}_{F_1,1}, \ldots, \mathsf{v}_{F_1,N_{d-1}^k} | \ldots | \mathsf{v}_{F_n,1}, \ldots, \mathsf{v}_{F_n,N_{d-1}^k} \right]^{\mathsf{T}}, \tag{8.4}$$

so that $\mathsf{v}_T \in \mathsf{V}_T^{k',k} := \mathbb{R}^{N_d^{k'}} \times (\mathbb{R}^{N_{d-1}^k})^n$. These coefficients are called *degrees of freedom* (DoFs). The structure of the array in (8.4) will guide us in the understanding of the structure of the matrices realizing the HHO operators.

Remark 8.3 (*p-refinement*) The setting can be generalized to account for different polynomial orders on each face $F_i \in \mathcal{F}_T$. This way, it becomes possible to use neighboring elements with different polynomial orders, opening the way to local p-refinement. The only required modification in the implementation is that the size of the sub-arrays in (8.4) needs to account for the different polynomial degrees. □

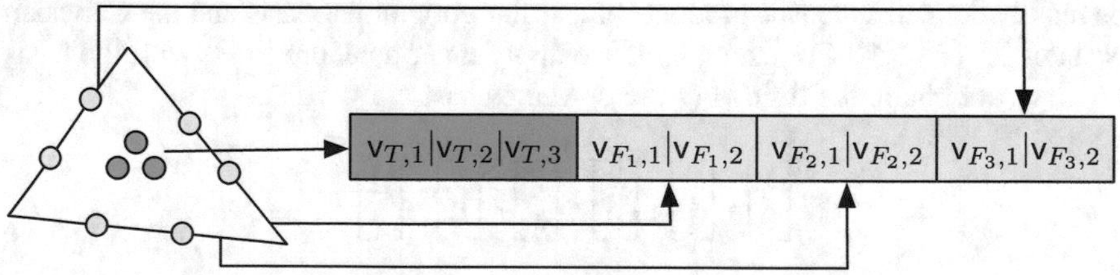

Fig. 8.2 Formation of the local vector of DoFs in the case of a triangle with $k' = k := 1$

8.3 L^2-Orthogonal Projections

L^2-orthogonal projections allow one to approximate functions belonging to a certain functional space with functions in a finite-dimensional polynomial space. Let us use a common notation $K \in \{T, F\}$ to denote a generic mesh cell or mesh face, with $d(T) := d$ and $d(F) := d - 1$. Given a function $v \in L^2(K)$, its projection $\Pi_K^k(v)$ on $\mathbb{P}_{d(K)}^k(K)$ is such that $(\Pi_K^k(v) - v, w)_{L^2(K)} = 0$ for all $w \in \mathbb{P}_{d(K)}^k(K)$. For notational convenience, let $p_K := \Pi_K^k(v)$. To compute p_K, we set up the problem

$$\int_K \left(\sum_{i \in \mathcal{N}_K^k} w_i \phi_{K,i}(\boldsymbol{x}) \sum_{j \in \mathcal{N}_K^k} p_{K,j} \phi_{K,j}(\boldsymbol{x}) \right) \mathrm{d}\boldsymbol{x} = \int_K \left(v(\boldsymbol{x}) \sum_{i \in \mathcal{N}_K^k} w_i \phi_{K,i}(\boldsymbol{x}) \right) \mathrm{d}\boldsymbol{x}, \tag{8.5}$$

where $\mathcal{N}_K^k := \{1, \ldots, N_{d(K)}^k\}$ and the functions $\{\phi_{K,i}\}_{i \in \mathcal{N}_K^k}$ are a set of basis functions attached to the geometric object K. By defining similarly the coefficient column vectors $\boldsymbol{p}_K = \{p_{K,i}\}_{i \in \mathcal{N}_K^k}$ and $\boldsymbol{w}$, and the basis function column vector $\boldsymbol{\phi}_K(\boldsymbol{x}) = \{\phi_{K,i}(\boldsymbol{x})\}_{i \in \mathcal{N}_K^k}$, the expression (8.5) can be rewritten in matrix form as

$$\boldsymbol{w}^{\mathsf{T}} \left(\int_K \boldsymbol{\phi}_K(\boldsymbol{x}) \boldsymbol{\phi}_K(\boldsymbol{x})^{\mathsf{T}} \, \mathrm{d}\boldsymbol{x} \right) \boldsymbol{p}_K = \boldsymbol{w}^{\mathsf{T}} \int_K v(\boldsymbol{x}) \boldsymbol{\phi}_K(\boldsymbol{x}) \, \mathrm{d}\boldsymbol{x}. \tag{8.6}$$

Since $p_K - v$ has to be orthogonal to all the test functions w, $\boldsymbol{p}_K$ is found by setting up and solving the linear system of $N_{d(K)}^k$ equations and $N_{d(K)}^k$ unknowns

$$\mathsf{M}_K \boldsymbol{p}_K = \int_K v(\boldsymbol{x}) \boldsymbol{\phi}_K(\boldsymbol{x}) \, \mathrm{d}\boldsymbol{x}, \tag{8.7}$$

with the mass matrix $\mathsf{M}_K := \int_K \boldsymbol{\phi}_K(\boldsymbol{x}) \boldsymbol{\phi}_K(\boldsymbol{x})^{\mathsf{T}} \, \mathrm{d}\boldsymbol{x}$ (by construction, M_K is symmetric positive-definite). One efficient way of solving the linear system (8.7) is to compute the Cholesky decomposition of M_K.

8.3.1 Quadratures

Integrals appearing in (8.6) are computed numerically using *quadrature rules*. A quadrature rule allows one to approximate integrals over the geometric element K as a weighted sum of evaluations of the integrand function f at certain points in K:

$$\int_K f(\boldsymbol{x}) \, \mathrm{d}\boldsymbol{x} \approx \sum_{q=1}^{|Q|} \omega_q f(\boldsymbol{x}_q), \tag{8.8}$$

where Q is a set composed of $|Q|$ pairs $(\boldsymbol{x}_q, \omega_q)$; for each pair, the first element is named *quadrature point*, and the second element is named *weight*. Quadratures are available for simplices, quadrilaterals, and hexahedra. These quadratures are conceived in a reference cell and mapped to the physical cell by an affine geometric mapping. Quadratures allow exact integration of polynomials up to a certain degree called the *quadrature order*. Integration on geometric objects having a more complex shape can be done by triangulating the geometric object and then employing a simplicial quadrature. Extensive literature about quadratures exists. Apart from the classical Gauss quadrature points [4], we mention [88, 100, 112] for quadratures on simplices and [51, 136, 137, 141] for quadratures on polygons and polyhedra based on various techniques that avoid the need to invoke a sub-triangulation.

By using the tools just introduced, the linear system (8.7) is set up numerically as

$$\left(\sum_{q=1}^{|Q|} \omega_q \boldsymbol{\phi}_K(\boldsymbol{x}_q) \boldsymbol{\phi}_K(\boldsymbol{x}_q)^{\mathsf{T}}\right) \boldsymbol{p}_K = \sum_{q=1}^{|Q|} \omega_q v(\boldsymbol{x}_q) \boldsymbol{\phi}_K(\boldsymbol{x}_q), \tag{8.9}$$

where Q needs to have a sufficient order to integrate exactly the product of the basis functions. For instance, if $\boldsymbol{\phi}_K$ is the vector of basis functions of $\mathbb{P}^k_{d(K)}(K)$, the quadrature needs to have the sufficient number of points to integrate exactly polynomials of degree $2k$.

8.3.2 Reduction Operator

Let $T \in \mathcal{T}$ be a mesh cell. The local HHO reduction operator can be rewritten in expanded form as

$$\hat{I}_T^{k',k}(v) := (\Pi_T^{k'}(v), \Pi_{F_1}^k(v_{|F_1}), \ldots, \Pi_{F_n}^k(v_{|F_n})) \in \hat{V}_T^{k',k}, \quad \forall v \in H^1(T). \tag{8.10}$$

The reduction is thus the collection of the projections of v on the cell T and on its n faces $F_1, \ldots, F_n$. At the algebraic level, this translates into obtaining the coefficients of $(n+1)$ polynomials by solving $(n+1)$ problems of the form (8.7). More precisely, let $\boldsymbol{\phi}_T$ be the vector of cell-based basis functions on the mesh cell T and let $\boldsymbol{\phi}_{F_i}$ be the vector of face-based basis functions on the i-th face of T. Moreover, let M_T and M_{F_i} be the corresponding mass matrices. The algebraic version of applying the reduction operator $\hat{I}_T^{k',k}$ to a function $v \in H^1(T)$ amounts to finding the array vector $\mathsf{I}_T^{k',k}(v) \in \mathsf{V}_T^{k',k}$ solving the following block-diagonal system:

$$\begin{bmatrix} \mathsf{M}_T & & & \\ & \mathsf{M}_{F_1} & & \\ & & \ddots & \\ & & & \mathsf{M}_{F_n} \end{bmatrix} \mathsf{I}_T^{k',k}(v) = \begin{bmatrix} \int_T v(x)\boldsymbol{\phi}_T(x)\,\mathrm{d}\boldsymbol{x} \\ \int_{F_1} v(x)\boldsymbol{\phi}_{F_1}(x)\,\mathrm{d}\boldsymbol{x} \\ \vdots \\ \int_{F_n} v(x)\boldsymbol{\phi}_{F_n}(x)\,\mathrm{d}\boldsymbol{x} \end{bmatrix}. \tag{8.11}$$

Even though it is not used in the actual HHO computations, the computation of $\mathsf{I}_T^{k',k}$ is essential to verify the correctness of the implementation of the reconstruction and stabilization operators detailed in the next section.

A possible implementation of the local reduction operator in 1D is shown in Listing 8.2. The function `hho_reduction()` takes the parameters `pd`, `elem` and `fun`, which are respectively a structure containing the computation parameters (in particular the polynomial degree and the cell diameter, which are taken here uniform on the whole mesh), the current element index, and the function to reduce. At line 7, we ask for a quadrature, obtaining the points, the weights and the size in the variables `qps`, `qws`, and `nn`, respectively. We then proceed with the `for` loop (line 10) building the mass matrix and the right-hand side; this loop corresponds to the summations in (8.9). The projection on the cell is finally computed at line 16 (in 1D, we just need to evaluate the function at the faces): compare the structure of the returned vector `I` with (8.4).

Listing 8.2 Possible implementation of the reduction operator in 1D.

```
1  % The HHO reduction operator
2  function I = hho_reduction(pd, elem, fun)
3      % pd.h: cell diameter, uniform for all elements
4      % pd.K: polynomial degree, equal for all elements
5      x_bar = cell_center(pd, elem);
6      [xF1, xF2] = face_centers(pd, elem);
7      [qps, qws, nn] = integrate(2*pd.K, pd.h, elem);
8      MM = zeros(pd.K+1, pd.K+1);
9      rhs = zeros(pd.K+1, 1);
10     for ii = 1:nn % This loop is the counterpart of (8.9)
11         [phi, ~] = basis(qps(ii), x_bar, pd.h, pd.K);
12         MM = MM + qws(ii) * (phi * phi');              % Mass matrix
13         rhs = rhs + qws(ii) * phi * fun(qps(ii));    % Right-hand side
14     end
15     I = zeros(pd.K+3, 1);
16     I(1:pd.K+1) = MM\rhs;    % Project on the cell
17     I(pd.K+2) = fun(xF1);     % Project on faces: in 1D we just need
18     I(pd.K+3) = fun(xF2);     %   to evaluate the function at the faces
19 end
```

Remark 8.4 (*Verifying the implementation*) Let us consider a sequence of successively refined meshes $\mathfrak{T} := (\mathcal{T}_i)_{i\in\mathbb{N}}$ and let h_i denote the maximum diameter of the cells composing $\mathcal{T}_i$. For each geometric object $K \in \{T, F\}$ of $\mathcal{T}_i \in \mathfrak{T}$, the projection on $\mathbb{P}^k_{d(K)}(K)$ of a function $v \in H^1(\Omega)$ is computed by solving the problem (8.7), obtaining a vector of DoFs $\boldsymbol{p}_K$. Such a vector is subsequently used to compute the global quantity

$$\epsilon_i := \left(\sum_K \int_K \left(v - \Pi_K^k(v)\right)^2 \mathrm{d}\boldsymbol{x}\right)^{1/2} = \left(\sum_K \sum_{q=1}^{|Q_K|} \omega_q \left(v(\boldsymbol{x}_q) - \boldsymbol{\phi}_K(\boldsymbol{x}_q)^{\mathsf{T}} \boldsymbol{p}_K\right)^2\right)^{1/2},$$

where Q_K is a quadrature of sufficient order on K and $\boldsymbol{\phi}_K$ is the vector of basis functions attached to K. The quantity ϵ_i has to decay, for increasing i, with rate $O(h_i^{k+1})$ if the summation is over the mesh cells, whereas it has to decay with a rate of $O(h_i^{k+1/2})$ if the summation is over the mesh faces (see Lemma 2.5). □

8.4 Algebraic Realization of the Local HHO Operators

Recalling Sect. 1.3, the reconstruction and stabilization operators lie at the heart of HHO methods. Both operators are locally defined in every mesh cell $T \in \mathcal{T}$ and map from the local HHO space $\hat{V}_T^{k',k}$ to some polynomial space: the reconstruction stabilization maps from $\hat{V}_T^{k',k}$ to $\mathbb{P}_d^{k+1}(T)$, whereas the stabilization operator restricted to each face $F \in \mathcal{F}_T$ maps from $\hat{V}_T^{k',k}$ to $\mathbb{P}_{d-1}^k(F)$. Since at the discrete level the elements of $\hat{V}_T^{k',k}$ translate to vectors of the form (8.4), both operators are represented by matrices that multiply a vector $\mathsf{v} \in \mathsf{V}_T^{k',k}$ to yield a vector representing either an element of $\mathbb{P}_d^{k+1}(T)$ or an element of $\mathbb{P}_{d-1}^k(F)$. This means that on a mesh cell with n faces, both matrices have $\hat{N}_d^{k',k} = N_d^{k'} + nN_{d-1}^k$ columns, which in turn form $n+1$ horizontally-juxtaposed blocks. We call T-block the first and leftmost block, whereas the remaining blocks are called F_i-blocks (see Fig. 8.3).

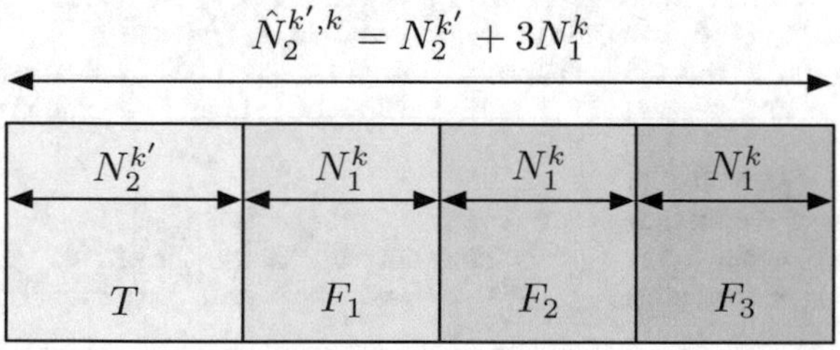

Fig. 8.3 Block structure of an HHO operator matrix on a triangle ($d = 2, n = 3$ faces). In particular, there is one T-block and three F_i-blocks, giving an horizontal size of $\hat{N}_2^{k',k} = N_2^{k'} + 3N_1^k$. The vertical size depends on the actual operator, as explained in the text

8.4.1 Local Reconstruction Operator

Let $T \in \mathcal{T}$. The local reconstruction operator satisfies (1.17), where we expand here the boundary term as a summation on the faces of the mesh cell as follows:

$$(\nabla R_T(\hat{v}_T), \nabla q)_{L^2(T)} = (\nabla v_T, \nabla q)_{L^2(T)} - \sum_{F \in \mathcal{F}_T} (v_T - v_F, \boldsymbol{n}_T{\cdot}\nabla q)_{L^2(F)}, \quad (8.12)$$

for all $q \in \mathbb{P}_d^{k+1}(T)^{\perp} := \{q \in \mathbb{P}_d^{k+1}(T) \mid (q, 1)_{L^2(T)} = 0\}$. Moreover, we have $(R_T(\hat{v}_T), 1)_{L^2(T)} = (v_T, 1)_{L^2(T)}$ (see (1.16)). It is however not necessary to work with the polynomial space $\mathbb{P}_d^{k+1}(T)^{\perp}$, and one can consider any subspace $\mathbb{P}_{*d}^{k+1}(T)$ leading to a direct sum $\mathbb{P}_d^{k+1}(T) = \mathbb{P}_d^0(T) \oplus \mathbb{P}_{*d}^{k+1}(T)$ (notice that $\dim(\mathbb{P}_{*d}^{k+1}(T)) = N_{*d}^{k+1} := N_d^{k+1} - 1$). One possibility is to consider basis functions of $\mathbb{P}_d^{k+1}(T)$ such that the first basis function is constant, and let the remaining basis functions span $\mathbb{P}_{*d}^{k+1}(T)$. Let $\boldsymbol{\varrho}(\boldsymbol{x})$ be the vector of basis functions of $\mathbb{P}_{*d}^{k+1}(T)$. Using a quadrature Q_T of order at least $2k$, the left-hand side of (8.12) is a plain stiffness matrix such that

$$\mathsf{K}_* := \sum_{q=1}^{|Q_T|} \omega_q \nabla\boldsymbol{\varrho}(\boldsymbol{x}_q) \cdot \nabla\boldsymbol{\varrho}(\boldsymbol{x}_q)^{\mathsf{T}}, \quad (8.13)$$

where ∇ is applied componentwise to $\boldsymbol{\varrho}(\boldsymbol{x})$ and the dot product only to the gradients. Notice that this computation results in a standard stiffness matrix, where the column and the row corresponding to the constant basis function have been dropped.

We next build the right-hand side of (8.12) in multiple steps. For simplicity, we assume that we are building the operator for a triangular element, so that $n := 3$ and $d := 2$. Let $\boldsymbol{\phi}_T(\boldsymbol{x})$ be the column vector of cell-based basis functions attached to T, $\boldsymbol{\phi}_{F_i}(\boldsymbol{x})$ the column vector of face-based basis functions attached to the face F_i (recall that these basis functions are computed using a geometric mapping from $\mathbb{R}^{d-1}$ to the hyperplane supporting F_i) and $\boldsymbol{0}_F$ a zero column vector of size N_{d-1}^k. We start with $(\nabla v_T, \nabla q)_{L^2(T)}$, where $v_T \in \mathbb{P}_d^{k'}(T)$ and $q \in \mathbb{P}_{*d}^{k+1}(T)$. In order to evaluate the cell-based part of a DoFs vector of the form (8.4), we form a column vector of basis functions $\boldsymbol{\mu}(\boldsymbol{x}) := [\boldsymbol{\phi}_T(\boldsymbol{x}) \mid \boldsymbol{0}_F \mid \boldsymbol{0}_F \mid \boldsymbol{0}_F\,]$, where $\mid$ denotes the vertical concatenation of column vectors. Then, we form the matrix

$$\mathsf{T} := \sum_{q=1}^{|Q_T|} \omega_q \nabla\boldsymbol{\varrho}(\boldsymbol{x}_q) \cdot \nabla\boldsymbol{\mu}(\boldsymbol{x}_q)^{\mathsf{T}}. \quad (8.14)$$

This computation yields a matrix where only the T-block has nonzero values. Its effect can be intuitively understood by looking separately at the roles of $\nabla\boldsymbol{\mu}(\boldsymbol{x}_q)^{\mathsf{T}}$ and $\nabla\boldsymbol{\varrho}(\boldsymbol{x}_q)$ when T multiplies a vector $\mathbf{v} \in \mathbf{V}_T^{k',k}$. For each quadrature point, $\boldsymbol{\mu}$ evaluates the gradients of the cell-based part of $\mathbf{v}$, whereas $\boldsymbol{\varrho}$ tests the value of the polynomial with the gradients of the basis functions of the reconstruction space.

In practice, this returns the right-hand side of a projection-like problem where the gradients of $\mathbb{P}^{k+1}_{*d}(T)$ are used as test functions.

We continue with the contributions from $(v_T - v_F, \boldsymbol{n}_T{\cdot}\nabla q)_{L^2(F)}$ for the $n = 3$ faces of T. For example, in order to compute the contribution due to the face F_1, we consider the vector of basis functions $\boldsymbol{\eta}_1(\boldsymbol{x}) := [\boldsymbol{\phi}_T(\boldsymbol{x}) \,|\, -\boldsymbol{\phi}_{F_1}(\boldsymbol{x}) \,|\, \mathbf{0}_F \,|\, \mathbf{0}_F\,]$. The contribution to the right-hand side is then computed as

$$\mathsf{F}_1 := \sum_{q_1=1}^{|Q_{F_1}|} \omega_{q_1} \boldsymbol{n}{\cdot}\nabla \boldsymbol{\varrho}(\boldsymbol{x}_{q_1}) \boldsymbol{\eta}_1(\boldsymbol{x}_{q_1})^{\mathsf{T}}. \tag{8.15}$$

Indeed, taking an array of the form (8.4) representing a member of $\hat{V}^{k',k}_T$ and computing the dot-product with $\boldsymbol{\eta}_1(\boldsymbol{x}_{q_1})$ corresponds to obtaining the value of the difference of the cell-based and F_1-based polynomials at the point $\boldsymbol{x}_{q_1} \in F_1$. Notice also that the matrix F_1 contains nonzero elements only in the T-block and in the F_1-block. The matrices F_2 and F_3 are computed in a similar fashion by taking $\boldsymbol{\eta}_2(\boldsymbol{x}) := [\boldsymbol{\phi}_T(\boldsymbol{x}) \,|\, \mathbf{0}_F \,|\, -\boldsymbol{\phi}_{F_2}(\boldsymbol{x}) \,|\, \mathbf{0}_F]$ and $\boldsymbol{\eta}_3(\boldsymbol{x}) := [\boldsymbol{\phi}_T(\boldsymbol{x}) \,|\, \mathbf{0}_F \,|\, \mathbf{0}_F \,|\, -\boldsymbol{\phi}_{F_3}(\boldsymbol{x})]$, respectively. If cells with more than three faces are used, the procedure is easily generalized by computing the remaining F_i matrices.

We finally compute the algebraic realization R of R_T (up to the mean-value constraint) by inverting the matrix K_* and setting

$$\mathsf{R} := \mathsf{K}_*^{-1}\mathsf{H} \quad \text{with} \quad \mathsf{H} := \mathsf{T} - \sum_{i=1}^{3} \mathsf{F}_i, \tag{8.16}$$

and the mean-value constraint can be satisfied by adding a suitable contribution from the constant basis function (and increasing by one the size of the vector R). Once we have computed R, we can readily obtain the matrix representing the stiffness term in (1.29) as

$$\mathsf{A} := \mathsf{R}^{\mathsf{T}}\mathsf{K}_*\mathsf{R} = \mathsf{H}^{\mathsf{T}}\mathsf{R}. \tag{8.17}$$

Take a moment to analyze the roles of the matrices composing A. K_* is a plain stiffness matrix on $\mathbb{P}^{k+1}_{*d}(T)$ and, as such, it operates on polynomials in $\mathbb{P}^{k+1}_{*d}(T)$ to compute a standard local stiffness term. In HHO however, DoFs live in the space $\mathbf{V}^{k',k}_T$: the reconstruction matrix R “translates” HHO DoFs to the higher-order space $\mathbb{P}^{k+1}_{*d}(T)$, on which K_* can operate. Listing 8.3 shows a possible realization of the computation of R in 1D. At lines 9-13, the stiffness matrix of $\mathbb{P}^{k+1}_d(T)$ is computed using a quadrature of order $2k$. It is subsequently trimmed to obtain K_* (line 16) and T (line 18). Starting from line 24, the boundary terms are computed. Finally, the reconstruction operator and the matrix A are obtained at lines 33 and 34, respectively. An illustration of the action of the reconstruction operator is shown in Fig. 8.4.

Listing 8.3 Possible implementation of the reconstruction operator in 1D.

```
1  % The HHO reconstruction operator
2  function [A, R] = hho_reconstruction(pd, elem)
3      x_bar = cell_center(pd, elem);
4      [xF1, xF2] = face_centers(pd, elem);
5
6      stiff_mat = zeros(pd.K+2, pd.K+2);
7      gr_rhs = zeros(pd.K+1, pd.K+3);
8
9      [qps, qws, nn] = integrate(2*pd.K, pd.h, elem);
10     for ii = 1:nn
11         [~, dphi] = basis(qps(ii), x_bar, pd.h, pd.K+1);
12         stiff_mat = stiff_mat + qws(ii) * (dphi * dphi');
13     end
14
15     % Set up local Neumann problem
16     gr_lhs = stiff_mat(2:end, 2:end); % Left-hand side
17     % Right-hand side, cell part
18     gr_rhs(:,1:pd.K+1) = stiff_mat(2:end,1:pd.K+1); % (∇ v_T, ∇ q)_{L^2(T)}
19
20     [phiF1, dphiF1] = basis(xF1, x_bar, pd.h, pd.K+1);
21     [phiF2, dphiF2] = basis(xF2, x_bar, pd.h, pd.K+1);
22
23     % Right-hand side, boundary part
24     gr_rhs(1:end, 1:pd.K+1) = gr_rhs(1:end, 1:pd.K+1) + ...
25         dphiF1(2:end)*phiF1(1:pd.K+1)'; % (v_T, n_T·∇ q)_{L^2(F_1)}
26
27     gr_rhs(1:end, 1:pd.K+1) = gr_rhs(1:end, 1:pd.K+1) - ...
28         dphiF2(2:end)*phiF2(1:pd.K+1)'; % (v_T, n_T·∇ q)_{L^2(F_2)}
29
30     gr_rhs(1:end, pd.K+2) = - dphiF1(2:end); % (v_F, n_T·∇ q)_{L^2(F_1)}
31     gr_rhs(1:end, pd.K+3) = + dphiF2(2:end); % (v_F, n_T·∇ q)_{L^2(F_2)}
32
33     R = gr_lhs\gr_rhs;      % Solve problem (up to a constant)
34     A = gr_rhs'*R;          % Compute (∇R_T(·), ∇R_T(·))_{L^2(T)}
35 end
```

Remark 8.5 (*Verifying the implementation*) Given a sequence of successively refined meshes $\mathfrak{T} = (\mathcal{T}_i)_{i\in\mathbb{N}}$ and a target function $v \in H^1(\Omega)$, the vector $\mathsf{I}_T^{k',k}(v) = [\mathsf{v}_T|\mathsf{v}_{F_1}|\dots|\mathsf{v}_{F_n}]^\mathsf{T}$ is computed for every mesh cell $T \in \mathcal{T}_i$ and all $i \in \mathbb{N}$. Subsequently, we compute the matrix-vector product $\mathsf{v}^* = \mathsf{R}\mathsf{I}_T^{k',k}(v)$, where v^* are the components of the polynomial $R_T(\hat{I}_T^{k',k}(v)) \in \mathbb{P}_{*d}^{k+1}(T)$. The average in T of the reconstructed function is fixed by forming the vector $\mathsf{v} = [v_\phi|\mathsf{v}^*]$, which collects the components of the reconstruction of v in $\mathbb{P}_d^{k+1}(T)$ and where v_ϕ is a constant ensuring the condition (1.16). We finally compute the L^2-error between the reconstruction of v and v itself, which should decay with rate $O(h_i^{k+2})$. □

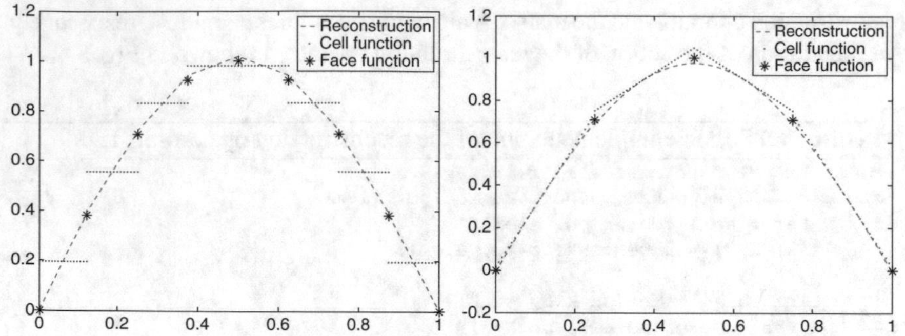

Fig. 8.4 Illustration of the action of the reconstruction operator R in 1D on the DoFs resulting from the computation of $\hat{I}_h^k(\sin(\pi x))$. On the left panel, the operator acts on cell polynomials of degree 0 (dotted line) and face values (stars) to reconstruct a piecewise polynomial of degree 1 (dashed line). On the right panel, starting from cell polynomials of degree 1 (dotted line) and face values (stars), a piecewise polynomial of degree 2 (dashed line) is reconstructed. Recall that in the 1D case, there is only one DoF per face

8.4.2 *The Stabilization Operator*

The computation of the HHO stabilization is a relatively involved task, and for this reason, it is discussed in two steps. In the first step, the Lehrenfeld–Schöberl (LS) stabilization is considered (recall that this stabilization is sufficient when working with mixed-order HHO methods, i.e., $k' = k + 1$). In the second step, the equal-order HHO stabilization is discussed as an extension of the LS stabilization.

Step 1: Lehrenfeld–Schöberl Stabilization

The idea behind the LS stabilization is to penalize just the difference between the polynomial attached to a face F_i and the trace on F_i of the polynomial attached to T. This is accomplished by using the operator $Z_F : \hat{V}_T^{k',k} \to \mathbb{P}_{d-1}^k(F)$ defined as

$$Z_F(\hat{v}_T) := \Pi_F^k(v_T) - v_F, \tag{8.18}$$

which is used to build the bilinear form $z_T : \hat{V}_T^{k',k} \times \hat{V}_T^{k',k} \to \mathbb{R}$ such that

$$z_T(\hat{v}_T, \hat{w}_T) := \sum_{F \in \mathcal{F}_T} h_T^{-1}(Z_F(\hat{v}_T), Z_F(\hat{w}_T))_{L^2(F)}. \tag{8.19}$$

The operator Z_F actually subtracts two polynomials, and at the algebraic level, this is done by subtracting their DoFs. This is accomplished by a matrix Z_i of size $N_{d-1}^k \times \hat{N}_d^k$ constructed as follows. A matrix I_i of size $N_{d-1}^k \times \hat{N}_d^k$ is first formed by placing a diagonal of ones in correspondence to the F_i-block (see Fig. 8.3) of I_i (note that the F_i-blocks of the stabilization operator are all square of size $N^k_{d-1} \times N^k_{d-1}$). The matrix I_i can be thought as a selection matrix such that when left-multiplying a vector of the form (8.4), it yields the subvector containing only the DoFs $\mathsf{v}_{F_i,1}, \ldots, \mathsf{v}_{F_i,N^k_{d-1}}$. The second step consists in computing the DoFs of the polynomial which represents the restriction on F_i of the polynomial attached to T, and this is done by means of a projection. We first construct the *trace matrix* of size $N^k_{d-1} \times \hat{N}^k_d$ such that

$$\mathsf{T}_i := \sum_{q=1}^{|Q_{F_i}|} \omega_q \boldsymbol{\phi}_{F_i}(\boldsymbol{x}_q)\boldsymbol{\mu}(\boldsymbol{x}_q)^{\mathsf{T}}, \tag{8.20}$$

whose role is explained as follows: for each quadrature point $\boldsymbol{x}_q \in Q_{F_i}$, if a vector of the form (8.4) is left-multiplied by $\boldsymbol{\mu}(\boldsymbol{x}_q)^{\mathsf{T}}$, the operation yields the value of the cell-based polynomial at the point $\boldsymbol{x}_q$ (which lies on F_i). The subsequent multiplication by $\boldsymbol{\phi}_{F_i}(\boldsymbol{x}_q)$ then tests the cell-based polynomial with the basis functions of $\mathbb{P}^k_{d-1}(F_i)$, effectively forming a right-hand side suitable for a projection problem like (8.6). The left-hand side of the projection problem is the mass matrix of the face F_i of size $N^k_{d-1} \times N^k_{d-1}$ such that

$$\mathsf{M}_i := \sum_{q=1}^{|Q_{F_i}|} \omega_q \boldsymbol{\phi}_{F_i}(\boldsymbol{x}_q)\boldsymbol{\phi}_{F_i}(\boldsymbol{x}_q)^{\mathsf{T}}, \tag{8.21}$$

with which we form the additional matrix $\mathsf{M}_i^{-1}\mathsf{T}_i$. This last matrix, applied to a vector of HHO DoFs, yields the sought restriction. Using the matrices just computed, we finally obtain the discrete counterpart of (8.18) as

$$\mathsf{Z}_i := \mathsf{M}_i^{-1}\mathsf{T}_i - \mathsf{I}_i, \tag{8.22}$$

which, if applied to a vector $\mathbf{v} \in \mathbf{V}_T^{k',k}$, yields the difference between the polynomial on F_i and the polynomial on T projected on the face F_i. This allows us to compute the algebraic counterpart of (8.19) as

$$\mathsf{Z} := \sum_{i=1}^{n} h_T^{-1}\mathsf{Z}_i^{\mathsf{T}}\mathsf{M}_i\mathsf{Z}_i. \tag{8.23}$$

Step 2: Equal-Order Stabilization

To obtain the equal-order HHO stabilization where $k' = k$, we need to enhance (8.18) by introducing a penalty on the high-order contribution due to the reconstruction. We consider (1.20), which we rewrite here by specifying the face $F \in \mathcal{F}_T$, leading to the operator $S_F : \hat{V}_T^k \to \mathbb{P}^k_{d-1}(F)$ defined as

$$S_F(\hat{v}_T) := Z_F(\hat{v}_T) + \Pi_F^k\big(R_T(\hat{v}_T) - \Pi_T^k R_T(\hat{v}_T)\big). \tag{8.24}$$

This operator is used to build the bilinear form $s_T : \hat{V}_T^k \times \hat{V}_T^k \to \mathbb{R}$ such that

$$s_T(\hat{v}_T, \hat{w}_T) := \sum_{F \in \mathcal{F}_T} h_T^{-1}(S_F(\hat{v}_T), S_F(\hat{w}_T))_{L^2(F)}. \tag{8.25}$$

We start by translating in matrix form the term $\Pi_F^k R_T(\hat{v}_T)$. First, we compute

$$\mathsf{T}_i' := \sum_{q=1}^{|Q_{F_i}|} \omega_q \boldsymbol{\phi}_{F_i}(\boldsymbol{x}_q)\boldsymbol{\varrho}(\boldsymbol{x}_q)^{\mathsf{T}}, \tag{8.26}$$

which has size $N_{d-1}^k \times N_{*d}^{k+1}$, to subsequently construct the term

$$\mathsf{M}_i^{-1}\mathsf{T}_i'\mathsf{R}, \tag{8.27}$$

where R is the reconstruction defined in (8.16) (notice that it is not necessary to take into account the mean-value correction in this construction). The matrix we just built can be understood by reading it backwards as follows: by applying R to an object in $\mathbf{V}_T^k$, we get its reconstruction in $\mathbb{P}_{*d}^{k+1}(T)$. Subsequently, when the trace matrix T_i' is applied to the DoFs of the reconstructed polynomial, its columns evaluate the DoFs of the reconstructed function on F_i, whereas the rows test it with the basis functions of $\mathbb{P}_{d-1}^k(F_i)$. The final multiplication by M_i^{-1} yields the DoFs of the reconstructed polynomial restricted to F_i.

We proceed similarly to translate the term $\Pi_F^k \Pi_T^k R_T(\hat{v}_T)$ in matrix form. This requires the introduction of the cell mass matrix M and the matrix

$$\mathsf{Q} := \sum_{q=1}^{|Q_T|} \omega_q \boldsymbol{\phi}_T(\boldsymbol{x}_q)\boldsymbol{\varrho}(\boldsymbol{x}_q)^{\mathsf{T}}, \tag{8.28}$$

which has size $N_d^k \times N_{*d}^{k+1}$. We construct the expression

$$\mathsf{M}_i^{-1}\widetilde{\mathsf{T}}_i\mathsf{M}^{-1}\mathsf{Q}\mathsf{R}, \tag{8.29}$$

where $\widetilde{\mathsf{T}}_i$ is the matrix T_i restricted to its first N_d^k columns. Again, this last expression is better understood by reading it backwards, and keeping in mind the role of the rows and the columns of each matrix: Q evaluates the DoFs of the reconstruction and tests it with the basis functions of $\mathbb{P}_d^k(T)$, whereas the multiplication by M^{-1} yields the DoFs corresponding to the result of the projection Π_T^k. The DoFs of the projection on the face are finally obtained by applying $\mathsf{M}_i^{-1}\widetilde{\mathsf{T}}_i$.

Putting everything together, the matrix form of the equal-order HHO stabilization is computed by combining (8.22), (8.27), and (8.29) as follows:

$$\mathsf{S}_i := \mathsf{Z}_i + \mathsf{M}_i^{-1}\mathsf{T}_i'\mathsf{R} - \mathsf{M}_i^{-1}\widetilde{\mathsf{T}}_i\mathsf{M}^{-1}\mathsf{QR}. \tag{8.30}$$

It is now possible to build the discrete counterpart of (8.25) as

$$\mathsf{S} := \sum_{i=1}^{n} h_T^{-1}\mathsf{S}_i^{\mathsf{T}}\mathsf{M}_i\mathsf{S}_i. \tag{8.31}$$

We propose in Listing 8.4 a practical implementation of the equal-order stabilization operator in 1D. On lines 7-8, the matrices `M` and `Q` are cut from an order $(k+1)$ mass matrix (`mass_mat`); an optimized construction would use an order k basis for the rows and a quadrature of order $2k+1$.

Listing 8.4 Possible implementation of the equal-order stabilization operator in 1D.

```
1  function S = hho_stabilization(pd, elem, R)
2      x_bar = cell_center(pd, elem);
3      [xF1, xF2] = face_centers(pd, elem);
4      mass_mat = make_mass_matrix(pd, elem, pd.K+1);
5
6      % Compute the term tmp1 = u_T - Π^k_T R_T(û_T)
7      M = mass_mat(1:pd.K+1,1:pd.K+1);
8      Q = mass_mat(1:pd.K+1,2:pd.K+2);
9      tmp1 = - M\(Q*R);
10     tmp1(1:pd.K+1, 1:pd.K+1) = tmp1(1:pd.K+1, 1:pd.K+1) + eye(pd.K+1);
11
12     [phiF1, ~] = basis(xF1, x_bar, pd.h, pd.K+1);
13     Mi = 1;
14     Ti = phiF1(2:end)';
15     Ti_tilde = phiF1(1:pd.K+1)';
16     tmp2 = Mi \ (Ti*R);                 % tmp2 = Π^k_F R_T(û_T)
17     tmp2(pd.K+2) = tmp2(pd.K+2)-1;      % tmp2 = Π^k_F R_T(û_T) - u_F
18     tmp3 = Mi \ (Ti_tilde * tmp1);      % tmp3 = Π^k_F(u_T - Π^k_T R_T(û_T))
19     Si = tmp2 + tmp3;                   % Si = Π^k_F R_T(û_T) - u_F + Π^k_F(u_T - Π^k_T R_T(û_T))
20     S = Si' * Mi * Si / pd.h;           % Accumulate on S
21
22     [phiF2, ~] = basis(xF2, x_bar, pd.h, pd.K+1);
23     Mi = 1;
24     Ti = phiF2(2:end)';
25     Ti_tilde = phiF2(1:pd.K+1)';
26     tmp2 = Mi \ (Ti*R);                 % tmp2 = Π^k_F(R_T(û_T))
27     tmp2(pd.K+3) = tmp2(pd.K+3)-1;      % tmp2 = Π^k_F R_T(û_T) - u_F
28     tmp3 = Mi \ (Ti_tilde * tmp1);      % tmp3 = Π^k_F(u_T - Π^k_T R_T(û_T))
29     Si = tmp2 + tmp3;                   % Si = Π^k_F R_T(û_T) - u_F + Π^k_F(u_T - Π^k_T R_T(û_T))
30     S = S + Si' * Mi * Si / pd.h;       % Accumulate on S
31 end
```

Remark 8.6 (*Verifying the implementation*) The correctness of the implementation of the stabilization operator is verified as before by taking a sequence of successively refined meshes $\mathfrak{T} = (\mathcal{T}_i)_{i\in\mathbb{N}}$ and a target function $v \in H^1(\Omega)$. For every mesh cell $T \in \mathcal{T}_i$ and all $i \in \mathbb{N}$, the local vector of DoFs $\mathbf{I}_T^{k',k}(v) = [\mathbf{v}_T|\mathbf{v}_{F_1}|\ldots|\mathbf{v}_{F_n}]^{\mathsf{T}}$ is computed. This vector is then used to compute the quantity $\epsilon_i := (\sum_{T\in\mathcal{T}_i} \mathbf{v}^{\mathsf{T}}\,\mathsf{S}\,\mathbf{v})^{\frac{1}{2}}$ which should converge to zero with decay rate $O(h_i^{k+1})$. The same result should be obtained for the LS stabilization. □

8.5 Assembly and Boundary Conditions

Using either the mixed-order or the equal-order HHO method, the local contributions in every mesh cell $T \in \mathcal{T}$ are computed as $\mathsf{L}_T := \mathsf{A}_T + \mathsf{Z}_T$ (using (8.17) and (8.23)) or $\mathsf{L}_T := \mathsf{A}_T + \mathsf{S}_T$ (using (8.17) and (8.31)), respectively. Here, we added a subscript referring to the mesh cell T for more clarity. The resulting local matrix L_T is statically condensed (see Sect. 1.4.2), leading to the condensed matrix $\mathsf{L}_T^{\mathrm{C}}$ of size $(nN_{d-1}^k) \times (nN_{d-1}^k)$ and the condensed right-hand side $\mathsf{b}_T^{\mathrm{C}}$ of size $(nN_{d-1}^k) \times 1$ (recall that $n := \#\mathcal{F}_T$ is the number of faces of T).

The assembly of the global problem requires a local-to-global correspondence array denoted by $\mathcal{G}_T : \{1, \ldots, n\} \to \{1, \ldots, \#\mathcal{F}\}$ for all $T \in \mathcal{T}$, between the local enumeration of the faces of T and their global enumeration as mesh faces. This array is usually provided by the mesh generator. In the first stage of the assembly process, one does not bother about boundary conditions (this amounts to assemble a problem with pure Neumann boundary conditions). The global matrix L^{G} is composed of $\#\mathcal{F} \times \#\mathcal{F}$ blocks of size $N_{d-1}^k \times N_{d-1}^k$ and the global right-hand side b^{G} is composed of $\#\mathcal{F}$ blocks of size $N_{d-1}^k \times 1$. Then the local contributions are assembled as follows: For all $T \in \mathcal{T}$,

$$\mathsf{L}^{\mathrm{G}}_{\mathcal{G}_T(i),\mathcal{G}_T(j)} \hookleftarrow \mathsf{L}^{\mathrm{C}}_{T;i,j} \quad \text{and} \quad \mathsf{b}^{\mathrm{G}}_{\mathcal{G}_T(i)} \hookleftarrow \mathsf{b}^{\mathrm{C}}_{T;i}, \qquad \forall i, j \in \{1, \ldots, n\}, \tag{8.32}$$

where we denote by $a \hookleftarrow b$ the operation of accumulating the value b on a, i.e. the statement `a = a + b` of the commonly used imperative programming languages. In other words, the local block (i, j) is summed to the global block in position $(\mathcal{G}_T(i), \mathcal{G}_T(j))$.

It remains to apply the boundary conditions. As discussed in [80], HHO methods can handle all the classical boundary conditions for the Poisson model problem (see also [37]). To fix the ideas, let us assume that the boundary is partitioned as $\partial\Omega = \overline{\partial\Omega_{\mathrm{N}}} \cup \overline{\partial\Omega_{\mathrm{D}}}$ leading to the following model problem:

$$-\Delta u = f \text{ in } \Omega, \qquad u = u_{\mathrm{D}} \text{ on } \partial\Omega_{\mathrm{D}}, \qquad \boldsymbol{n}{\cdot}\nabla u = g_{\mathrm{N}} \text{ on } \partial\Omega_{\mathrm{N}}. \tag{8.33}$$

We assume that every mesh boundary face belongs either to $\partial\Omega_{\mathrm{D}}$ or to $\partial\Omega_{\mathrm{N}}$; the corresponding subsets of $\mathcal{F}^{\partial}$ are denoted by $\mathcal{F}^{\partial}_{\mathrm{D}}$ and $\mathcal{F}^{\partial}_{\mathrm{N}}$. Let us consider an idealized 1D situation with a simple mesh containing only four faces (vertices), i.e., $\mathcal{F} := \{F_1, F_2, F_3, F_4\}$ with $\mathcal{F}^{\circ} := \{F_2, F_3\}$ and $\mathcal{F}^{\partial} := \{F_1, F_4\}$. Then, assuming that only Neumann boundary conditions are enforced (i.e., $\partial\Omega = \partial\Omega_{\mathrm{N}}$, $\partial\Omega_{\mathrm{D}} = \emptyset$ in (8.33)), the global problem takes the form

$$\begin{bmatrix} \mathsf{L}_{11} & \mathsf{L}_{12} & & \\ \mathsf{L}_{21} & \mathsf{L}_{22} & \mathsf{L}_{23} & \\ & \mathsf{L}_{32} & \mathsf{L}_{33} & \mathsf{L}_{34} \\ & & \mathsf{L}_{43} & \mathsf{L}_{44} \end{bmatrix} \begin{bmatrix} \mathbf{u}_1 \\ \mathbf{u}_2 \\ \mathbf{u}_3 \\ \mathbf{u}_4 \end{bmatrix} = \begin{bmatrix} \boldsymbol{b}_1 \\ \boldsymbol{b}_2 \\ \boldsymbol{b}_3 \\ \boldsymbol{b}_4 \end{bmatrix}. \tag{8.34}$$

(Notice that in this 1D case, all the entries are actually scalars.) Assume now that the Neumann boundary condition is applied only on F_1 and that the Dirichlet condition is applied on F_4. Then we have $\mathbf{u}_4 = \mathsf{M}_{F_4}^{-1}\boldsymbol{d}_4$ with $\boldsymbol{d}_4 := \int_{F_4} u_{\mathrm{D}}(\boldsymbol{x})\boldsymbol{\phi}_{F_4}(\boldsymbol{x})\,\mathrm{d}s$. Eliminating $\mathbf{u}_4$ from the first three rows of (8.34) gives the reduced system

$$\begin{bmatrix} \mathsf{L}_{11} & \mathsf{L}_{12} & \\ \mathsf{L}_{21} & \mathsf{L}_{22} & \mathsf{L}_{23} \\ & \mathsf{L}_{32} & \mathsf{L}_{33} \end{bmatrix} \begin{bmatrix} \mathbf{u}_1 \\ \mathbf{u}_2 \\ \mathbf{u}_3 \end{bmatrix} = \begin{bmatrix} \boldsymbol{b}_1 \\ \boldsymbol{b}_2 \\ \boldsymbol{b}_3' \end{bmatrix}, \tag{8.35}$$

where $\boldsymbol{b}_3' := \boldsymbol{b}_3 - \mathsf{L}_{34}\mathsf{M}_{F_4}^{-1}\boldsymbol{d}_4$. This process can be conveniently done on the fly during the assembly, but a new mapping $\mathcal{G}^{\circ}$ has to be used. Such a mapping is computed like $\mathcal{G}$, but removing the Dirichlet faces. Once the solution of the reduced system is found, the full solution is recovered by plugging $\mathbf{d}_4$ after $\mathbf{u}_3$ in the solution vector. An alternative approach is to introduce a Lagrange multiplier to enforce the Dirichlet condition:

$$\begin{bmatrix} \mathsf{L}_{11} & \mathsf{L}_{12} & & & \\ \mathsf{L}_{21} & \mathsf{L}_{22} & \mathsf{L}_{23} & & \\ & \mathsf{L}_{32} & \mathsf{L}_{33} & \mathsf{L}_{34} & \\ & & \mathsf{L}_{43} & \mathsf{L}_{44} & \mathsf{M}_{F_4} \\ & & & \mathsf{M}_{F_4} & \end{bmatrix} \begin{bmatrix} \mathbf{u}_1 \\ \mathbf{u}_2 \\ \mathbf{u}_3 \\ \mathbf{u}_4 \\ \boldsymbol{\lambda}_5 \end{bmatrix} = \begin{bmatrix} \boldsymbol{b}_1 \\ \boldsymbol{b}_2 \\ \boldsymbol{b}_3 \\ \boldsymbol{b}_4 \\ \boldsymbol{d}_4 \end{bmatrix}. \tag{8.36}$$

This second technique leads to a slightly larger system having a saddle-point structure, but it could be easier to implement in a first HHO code.

Remark 8.7 (*Neumann boundary conditions*) The Neumann boundary condition in the model problem (8.33) leads to a modification of the linear form on the right-hand side of the discrete problem, which reads $\ell(\hat{w}_h) := (f, w_{\mathcal{T}})_{L^2(\Omega)} + (g_{\mathrm{N}}, w_{\mathcal{F}})_{L^2(\partial\Omega_{\mathrm{N}})}$ (see Sect. 4.2.2 for the linear elasticity problem). At the implementation level, the Neumann condition reduces to a contribution on the right-hand side of the linear system positioned according to the Neumann face unknowns. Such a contribution is computed as $\boldsymbol{g}_i = \sum_{q=1}^{|Q_{F_i}|} \omega_q g_{\mathrm{N}}(\boldsymbol{x}_q)\boldsymbol{\phi}_{F_i}(\boldsymbol{x}_q)$, where $\boldsymbol{\phi}_{F_i}$ is the vector of basis functions of the globally-numbered i-th face. Those contributions are then added to the i-th block of the right-hand side. □

8.6 Remarks on the Computational Cost of HHO Methods

The computational costs in HHO methods are of two kinds: local costs associated with the assembly and global costs associated with the solution of the global linear system. We focus as before on the Poisson model problem.

The local costs include the computation of the operators and the static condensation, and they differ in the mixed-order and equal-order methods. In the mixed-order method, the computation of the reconstruction and the static condensation are slightly more expensive compared to the equal-order case, essentially because of the increased number of cell-based DoFs. The costs of the stabilization, however, differ substantially between the two variants of the method. This fact can be deduced by comparing the structure of (8.22) and (8.30). The mixed-order stabilization requires n inversions of the face mass matrices M_i, for a cost of $n \cdot O((N_{d-1}^k)^3) \approx O(k^{3d-3})$, together with the construction of the trace matrices T_i, for a cost of $n \cdot O((k+1)^{d-1} \cdot N_d^{k+1} \cdot N_{d-1}^k) \approx O(k^{3d-2})$. Instead, the equal-order stabilization requires the inversion of the cell mass matrix and other operations which are at least cubic in the size of the cell basis. To illustrate this fact, we performed some computational experiments on common element types, namely triangles and quadrangles in 2D, and tetrahedra and hexahedra in 3D (see Fig. 8.5). In all cases, we

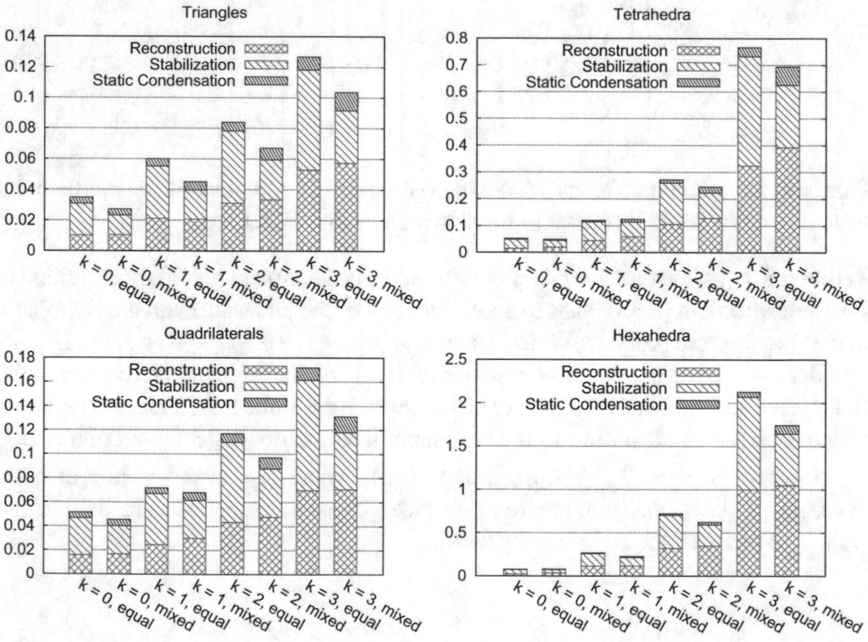

Fig. 8.5 Comparison of average computational times (in milliseconds) for the construction of the HHO operators on a single mesh cell, including static condensation. In the mixed-order HHO method, even if the cost of reconstruction and static condensation is slightly increased, the lower cost of stabilization results in a reduction of total computational time

Table 8.1 Comparative cost assessment between HHO and SIP-DG on a tetrahedral mesh composed of 3,072 elements

	HHO(k, k)				SIP-DG$(k+1)$			
k	L^2-error	DoFs	Mflops	Memory (MB)	L^2-error	DoFs	Mflops	Memory (MB)
0	1.73e–2	5760	38	39	2.14e–2	12,288	787	85
1	1.06e–3	17,280	1006	106	4.61e–4	30,720	11,429	319
2	9.05e–5	34,560	8723	292	2.14e–5	61,440	92,799	1108
3	6.45e–6	57,600	40,389	719	1.04e–6	107,520	497,245	3215

Table 8.2 Comparative cost assessment between HHO and SIP-DG on a hexahedral mesh composed of 4096 ($16 \times 16 \times 16$) elements

	HHO(k, k)				SIP-DG$(k+1)$			
k	L^2-error	DoFs	Mflops	Memory (MB)	L^2-error	DoFs	Mflops	Memory (MB)
0	9.07e–3	11,520	310	64	6.03e–2	16,384	6677	168
1	3.04e–4	34,560	9671	293	1.72e–4	40,960	104,199	765
2	1.73e–5	69,120	58,977	884	1.29e–6	81,920	845,545	2844
3	7.13e–7	115,200	349,664	2412	5.24e–8	143,360	4,592,328	8490

observe that when using the mixed-order HHO method, even if one pays a bit more in reconstruction and static condensation, one pays a lot less in stabilization. This turns in an overall reduction of the cost of the computation of the local contributions.

Concerning the global costs, we illustrate the differences between the equal-order HHO method and the well-established symmetric interior-penalty discontinuous Galerkin (SIP-DG) method (see [8] or [76, Sect. 4.2]) for the Poisson model problem posed in the unit cube $(0, 1)^3$. For HHO, we use polynomials of one degree less than in SIP-DG, so that both methods deliver the same error decay rates. We ran the experiments on 3D meshes of tetrahedra (3,072 elements) and hexahedra (4,096 elements). The global linear systems were solved using the PARDISO linear solver from the Intel MKL library. Memory usage was estimated via the `getrusage()` system call. The results reported in Tables 8.1 and 8.2 indicate that the HHO discretization is more favorable in terms of linear solver operations and memory usage.

References

1. M. Abbas, A. Ern, and N. Pignet. Hybrid High-Order methods for finite deformations of hyperelastic materials. *Comput. Mech.*, 62(4):909–928, 2018.
2. M. Abbas, A. Ern, and N. Pignet. A Hybrid High-Order method for incremental associative plasticity with small deformations. *Comput. Methods Appl. Mech. Engrg.*, 346:891–912, 2019.
3. M. Abbas, A. Ern, and N. Pignet. A Hybrid High-Order method for finite elastoplastic deformations within a logarithmic strain framework. *Internat. J. Numer. Methods Engrg.*, 120(3):303–327, 2019.
4. M. Abramowitz and I. A. Stegun. *Handbook of Mathematical Functions with Formulas, Graphs, and Mathematical Tables*. Dover, New York, NY, 1972.
5. R. A. Adams and J. J. F. Fournier. *Sobolev spaces*, volume 140 of *Pure and Applied Mathematics (Amsterdam)*. Elsevier/Academic Press, Amsterdam, second edition, 2003.
6. J. Aghili, S. Boyaval, and D. A. Di Pietro. Hybridization of mixed high-order methods on general meshes and application to the Stokes equations. *Comput. Methods Appl. Math.*, 15(2):111–134, 2015.
7. D. Anderson and J. Droniou. An arbitrary-order scheme on generic meshes for miscible displacements in porous media. *SIAM J. Sci. Comput.*, 40(4):B1020–B1054, 2018.
8. D. N. Arnold, F. Brezzi, B. Cockburn, and L. D. Marini. Unified analysis of discontinuous Galerkin methods for elliptic problems. *SIAM J. Numer. Anal.*, 39(5):1749–1779, 2001/02.
9. E. Artioli, L. Beirão da Veiga, C. Lovadina, and E. Sacco. Arbitrary order 2D virtual elements for polygonal meshes: part II, inelastic problem. *Comput. Mech.*, 60(4):643–657, 2017.
10. B. Ayuso de Dios, K. Lipnikov, and G. Manzini. The nonconforming virtual element method. *ESAIM Math. Model. Numer. Anal.*, 50(3):879–904, 2016.
11. G. A. Baker. Error estimates for finite element methods for second order hyperbolic equations. *SIAM J. Numer. Anal.*, 13(4):564–576, 1976.
12. J. M. Ball. Convexity conditions and existence theorems in nonlinear elasticity. *Arch. Rational Mech. Anal.*, 63(4):337–403, 1976/77.
13. M. Bebendorf. A note on the Poincaré inequality for convex domains. *Z. Anal. Anwendungen*, 22(4):751–756, 2003.
14. L. Beirão da Veiga, F. Brezzi, A. Cangiani, G. Manzini, L. D. Marini, and A. Russo. Basic principles of virtual element methods. *M3AS Math. Models Methods Appl. Sci.*, 199(23):199–214, 2013.

M. Cicuttin et al., *Hybrid High-Order Methods*,
SpringerBriefs in Mathematics,
https://doi.org/10.1007/978-3-030-81477-9

15. L. Beirão da Veiga, C. Lovadina, and D. Mora. A virtual element method for elastic and inelastic problems on polytope meshes. *Comput. Methods Appl. Mech. Engrg.*, 295:327–346, 2015.
16. D. Boffi, M. Botti, and D. A. Di Pietro. A nonconforming high-order method for the Biot problem on general meshes. *SIAM J. Sci. Comput.*, 38(3):A1508–A1537, 2016.
17. D. Boffi, F. Brezzi, and M. Fortin. *Mixed finite element methods and applications*, volume 44 of *Springer Series in Computational Mathematics*. Springer, Heidelberg, 2013.
18. L. Boillot. *Contributions to the mathematical modeling and to the parallel algorithmic for the optimization of an elastic wave propagator in anisotropic media*. PhD thesis, Université de Pau et des Pays de l'Adour, France, 2014.
19. F. Bonaldi, D. A. Di Pietro, G. Geymonat, and F. Krasucki. A Hybrid High-Order method for Kirchhoff-Love plate bending problems. *ESAIM Math. Model. Numer. Anal.*, 52(2):393–421, 2018.
20. J. Bonet and R. D Wood. *Nonlinear continuum mechanics for finite element analysis*. Cambridge University Press, Cambridge, 1997.
21. L. Botti and D. A. Di Pietro. Assessment of Hybrid High-Order methods on curved meshes and comparison with discontinuous Galerkin methods. *J. Comput. Phys.*, 370:58–84, 2018.
22. L. Botti, D. A. Di Pietro, and J. Droniou. A Hybrid High-Order discretisation of the Brinkman problem robust in the Darcy and Stokes limits. *Comput. Methods Appl. Mech. Engrg.*, 341:278–310, 2018.
23. L. Botti, D. A. Di Pietro, and J. Droniou. A Hybrid High-Order method for the incompressible Navier-Stokes equations based on Temam's device. *J. Comput. Phys.*, 376:786–816, 2019.
24. M. Botti, D. Castañón Quiroz, D. A. Di Pietro, and A. Harnist. A Hybrid High-Order method for creeping flows of non-Newtonian fluids. *ESAIM Math. Model. Numer. Anal.*, 55(5):2045–2073, 2021.
25. M. Botti, D. A. Di Pietro, and A. Guglielmana. A low-order nonconforming method for linear elasticity on general meshes. *Comput. Methods Appl. Mech. Engrg.*, 354:96–118, 2019.
26. M. Botti, D. A. Di Pietro, and P. Sochala. A Hybrid High-Order method for nonlinear elasticity. *SIAM J. Numer. Anal.*, 55(6):2687–2717, 2017.
27. M. Botti, D. A. Di Pietro, and P. Sochala. A hybrid high-order discretization method for nonlinear poroelasticity. *Comput. Methods Appl. Math.*, 20(2):227–249, 2020.
28. S. C. Brenner. Poincaré-Friedrichs inequalities for piecewise H^1 functions. *SIAM J. Numer. Anal.*, 41(1):306–324, 2003.
29. H. Brezis. Équations et inéquations non linéaires dans les espaces vectoriels en dualité. *Ann. Inst. Fourier (Grenoble)*, 18(fasc. 1):115–175, 1968.
30. H. Brezis. *Functional analysis, Sobolev spaces and partial differential equations*. Universitext. Springer, New York, NY, 2011.
31. E. Burman, M. Cicuttin, G. Delay, and A. Ern. An unfitted hybrid high-order method with cell agglomeration for elliptic interface problems. *SIAM J. Sci. Comput.*, 43(2):A859–A882, 2021.
32. E. Burman, G. Delay, and A. Ern. An unfitted hybrid high-order method for the Stokes interface problem. *IMA J. Numer. Anal.*, 41(4):2362–2387, 2021.
33. E. Burman, O. Duran, and A. Ern. Hybrid high-order methods for the acoustic wave equation in the time domain. *Commun. Appl. Math. Comput.*, 2021. hal-02922702.
34. E. Burman, O. Duran, A. Ern, and M. Steins. Convergence analysis of hybrid high-order methods for the wave equation. *J. Sci. Comput.*, 87(3):Paper No. 91, 30, 2021.
35. E. Burman and A. Ern. An unfitted hybrid high-order method for elliptic interface problems. *SIAM J. Numer. Anal.*, 56(3):1525–1546, 2018.
36. E. Burman and A. Ern. A cut cell hybrid high-order method for elliptic problems with curved boundaries. In *Numerical mathematics and advanced applications—ENUMATH 2017*, volume 126 of *Lect. Notes Comput. Sci. Eng.*, pages 173–181. Springer, Cham, 2019.
37. R. Bustinza and J. Munguia-La-Cotera. A hybrid high-order formulation for a Neumann problem on polytopal meshes. *Numer. Methods Partial Differential Equations*, 36(3):524–551, 2020.

38. V. Calo, M. Cicuttin, Q. Deng, and A. Ern. Spectral approximation of elliptic operators by the hybrid high-order method. *Math. Comp.*, 88(318):1559–1586, 2019.
39. A. Cangiani, Z. Dong, and E. H. Georgoulis. *hp*-version discontinuous Galerkin methods on essentially arbitrarily-shaped elements. *Math. Comp.*, published online, 2021. arXiv preprint 1906.01715.
40. A. Cangiani, Z. Dong, E. H. Georgoulis, and P. Houston. *hp*-version discontinuous Galerkin methods on polygonal and polyhedral meshes. Springer Briefs in Mathematics. Springer, Cham, 2017.
41. C. Carstensen, A. Ern, and S. Puttkammer. Guaranteed lower bounds on eigenvalues of elliptic operators with a hybrid high-order method. *Numer. Math.*, 149(2):273–304, 2021.
42. C. Carstensen and S. A. Funken. Constants in Clément-interpolation error and residual based a posteriori error estimates in finite element methods. *East-West J. Numer. Math.*, 8(3):153–175, 2000.
43. K. L. Cascavita, J. Bleyer, X. Chateau, and A. Ern. Hybrid discretization methods with adaptive yield surface detection for Bingham pipe flows. *J. Sci. Comput.*, 77(3):1424–1443, 2018.
44. K. L. Cascavita, F. Chouly, and A. Ern. Hybrid high-order discretizations combined with Nitsche's method for Dirichlet and Signorini boundary conditions. *IMA J. Numer. Anal.*, 40(4):2189–2226, 2020.
45. D. Castañón Quiroz and D. A. Di Pietro. A Hybrid High-Order method for the incompressible Navier–Stokes problem robust for large irrotational body forces. *Comput. Math. Appl.*, 79(8):2655–2677, 2020.
46. T. Chaumont-Frelet, A. Ern, S. Lemaire, and F. Valentin. Bridging the multiscale hybrid-mixed and multiscale hybrid high-order methods. hal-03235525, 2021.
47. F. Chave, D. A. Di Pietro, and L. Formaggia. A Hybrid High-Order method for Darcy flows in fractured porous media. *SIAM J. Sci. Comput.*, 40(2):1063–1094, 2018.
48. F. Chave, D. A. Di Pietro, and S. Lemaire. A discrete Weber inequality on three-dimensional hybrid spaces with application to the HHO approximation of magnetostatics. hal-02892526, 2020.
49. F. Chave, D. A. Di Pietro, F. Marche, and F. Pigeonneau. A Hybrid High-Order method for the Cahn-Hilliard problem in mixed form. *SIAM J. Numer. Anal.*, 54(3):1873–1898, 2016.
50. H. Chi, L. Beirão da Veiga, and G. H. Paulino. Some basic formulations of the virtual element method (VEM) for finite deformations. *Comput. Methods Appl. Mech. Engrg.*, 318:148–192, 2017.
51. E. B. Chin, J. B. Lasserre, and N. Sukumar. Numerical integration of homogeneous functions on convex and nonconvex polygons and polyhedra. *Comput. Mech.*, 56(6):967–981, 2015.
52. F. Chouly. An adaptation of Nitsche's method to the Tresca friction problem. *J. Math. Anal. Appl.*, 411:329–339, 2014.
53. F. Chouly, A. Ern, and N. Pignet. A hybrid high-order discretization combined with Nitsche's method for contact and Tresca friction in small strain elasticity. *SIAM J. Sci. Comput.*, 42(4):A2300–A2324, 2020.
54. F. Chouly and P. Hild. A Nitsche-based method for unilateral contact problems: numerical analysis. *SIAM J. Numer. Anal.*, 51(2):1295–1307, 2013.
55. F. Chouly, P. Hild, and Y. Renard. Symmetric and non-symmetric variants of Nitsche's method for contact problems in elasticity: theory and numerical experiments. *Math. Comp.*, 84(293):1089–1112, 2015.
56. P. G. Ciarlet. *Mathematical elasticity. Vol. I*, volume 20 of *Studies in Mathematics and its Applications*. North-Holland Publishing Co., Amsterdam, 1988.
57. P. G. Ciarlet. *The finite element method for elliptic problems*, volume 40 of *Classics in Applied Mathematics*. Society for Industrial and Applied Mathematics (SIAM), Philadelphia, PA, 2002. Reprint of the 1978 original [North-Holland, Amsterdam].
58. M. Cicuttin, D. A. Di Pietro, and A. Ern. Implementation of Discontinuous Skeletal methods on arbitrary-dimensional, polytopal meshes using generic programming. *J. Comput. Appl. Math.*, 344:852–874, 2018.

59. M. Cicuttin, A. Ern, and T. Gudi. Hybrid high-order methods for the elliptic obstacle problem. *J. Sci. Comput.*, 83(1):Paper No. 8, 18, 2020.
60. M. Cicuttin, A. Ern, and S. Lemaire. A Hybrid High-Order method for highly oscillatory elliptic problems. *Comput. Methods Appl. Math.*, 19(4):723–748, 2019.
61. B. Cockburn. Static condensation, hybridization, and the devising of the HDG methods. In G. R. Barrenechea, F. Brezzi, A. Cangiani, and E. H. Georgoulis, editors, *Building Bridges: Connections and Challenges in Modern Approaches to Numerical Partial Differential Equations*, volume 114 of *Lecture Notes in Computational Science and Engineering*, pages 129–178, Springer, Cham, 2016.
62. B. Cockburn, D. A. Di Pietro, and A. Ern. Bridging the Hybrid High-Order and hybridizable discontinuous Galerkin methods. *ESAIM Math. Model. Numer. Anal.*, 50(3):635–650, 2016.
63. B. Cockburn, Z. Fu, A. Hungria, L. Ji, M. A. Sánchez, and F.-J. Sayas. Störmer-Numerov HDG methods for acoustic waves. *J. Sci. Comput.*, 75(2):597–624, 2018.
64. B. Cockburn, J. Gopalakrishnan, and R. Lazarov. Unified hybridization of discontinuous Galerkin, mixed, and continuous Galerkin methods for second order elliptic problems. *SIAM J. Numer. Anal.*, 47(2):1319–1365, 2009.
65. B. Cockburn, J. Gopalakrishnan, and F.-J. Sayas. A projection-based error analysis of HDG methods. *Math. Comp.*, 79(271):1351–1367, 2010.
66. B. Cockburn, W. Qiu, and K. Shi. Conditions for superconvergence of HDG methods for second-order elliptic problems. *Math. Comp.*, 81(279):1327–1353, 2012.
67. B. Cockburn and V. Quenneville-Bélair. Uniform-in-time superconvergence of the HDG methods for the acoustic wave equation. *Math. Comp.*, 83(285):65–85, 2014.
68. A. Curnier and P. Alart. A generalized Newton method for contact problems with friction. *J. Méc. Théor. Appl.*, 7(suppl. 1):67–82, 1988.
69. J. Dabaghi and G. Delay. A unified framework for high-order numerical discretizations of variational inequalities. hal-02969793, 2020.
70. G. Dal Maso, A. DeSimone, and M. G. Mora. Quasistatic evolution problems for linearly elastic–perfectly plastic materials. *Arch. Ration. Mech. Anal.*, 180(2):237–291, 2006.
71. M. Dauge. *Elliptic boundary value problems on corner domains*, volume 1341 of *Lecture Notes in Mathematics*. Springer-Verlag, Berlin, 1988.
72. D. A. Di Pietro and J. Droniou. A Hybrid High-Order method for Leray-Lions elliptic equations on general meshes. *Math. Comp.*, 86(307):2159–2191, 2017.
73. D. A. Di Pietro and J. Droniou. *The Hybrid High-Order method for polytopal meshes*, volume 19 of *Modeling, Simulation and Application*. Springer, Cham, 2020.
74. D. A. Di Pietro, J. Droniou, and A. Ern. A discontinuous-skeletal method for advection-diffusion-reaction on general meshes. *SIAM J. Numer. Anal.*, 53(5):2135–2157, 2015.
75. D. A. Di Pietro, J. Droniou, and G. Manzini. Discontinuous Skeletal Gradient Discretisation Methods on polytopal meshes. *J. Comput. Phys.*, 355:397–425, 2018.
76. D. A. Di Pietro and A. Ern. *Mathematical Aspects of Discontinuous Galerkin Methods*, volume 69 of *Mathématiques & Applications*. Springer-Verlag, Berlin, 2012.
77. D. A. Di Pietro and A. Ern. A Hybrid High-Order locking-free method for linear elasticity on general meshes. *Comput. Meth. Appl. Mech. Engrg.*, 283:1–21, 2015.
78. D. A. Di Pietro and A. Ern. Arbitrary-order mixed methods for heterogeneous anisotropic diffusion on general meshes. *IMA J. Numer. Anal.*, 37(1):40–63, 2017. Preprint originally available at hal-00918482v1 (2013).
79. D. A. Di Pietro, A. Ern, and S. Lemaire. An arbitrary-order and compact-stencil discretization of diffusion on general meshes based on local reconstruction operators. *Comput. Meth. Appl. Math.*, 14(4):461–472, 2014.
80. D. A. Di Pietro, A. Ern, and S. Lemaire. A review of Hybrid High-Order methods: formulations, computational aspects, comparison with other methods. In *Building bridges: connections and challenges in modern approaches to numerical partial differential equations*, volume 114 of *Lect. Notes Comput. Sci. Eng.*, pages 205–236. Springer, Cham, 2016.
81. D. A. Di Pietro, A. Ern, A. Linke, and F. Schieweck. A discontinuous skeletal method for the viscosity-dependent Stokes problem. *Comput. Methods Appl. Mech. Engrg.*, 306:175–195, 2016.

82. D. A. Di Pietro and S. Krell. A Hybrid High-Order method for the steady incompressible Navier–Stokes problem. *J. Sci. Comput.*, 74(3):1677–1705, 2018.
83. D. A. Di Pietro and R. Specogna. An a posteriori-driven adaptive mixed high-order method with application to electrostatics. *J. Comput. Phys.*, 326:35–55, 2016.
84. J. K. Djoko, F. Ebobisse, A. T. McBride, and B. D. Reddy. A discontinuous Galerkin formulation for classical and gradient plasticity. I. Formulation and analysis. *Comput. Methods Appl. Mech. Engrg.*, 196(37-40):3881–3897, 2007.
85. J. Droniou, R. Eymard, T. Gallouët, and R. Herbin. A unified approach to mimetic finite difference, hybrid finite volume and mixed finite volume methods. *Math. Models Methods Appl. Sci.*, 20(2):265–295, 2010.
86. J. Droniou and B. P. Lamichhane. Gradient schemes for linear and non-linear elasticity equations. *Numer. Math.*, 129(2):251–277, 2015.
87. S. Du and F.-J. Sayas. *An invitation to the theory of the hybridizable discontinuous Galerkin method*. Springer Briefs in Mathematics. Springer, Cham, 2019.
88. D. A. Dunavant. High degree efficient symmetrical Gaussian quadrature rules for the triangle. *Internat. J. Numer. Methods Engrg.*, 21(6):1129–1148, 1985.
89. T. Dupont. L^2-estimates for Galerkin methods for second order hyperbolic equations. *SIAM J. Numer. Anal.*, 10:880–889, 1973.
90. Electricité de France. Finite element code_aster, structures and thermomechanics analysis for studies and research. Open source on www.code-aster.org, 1989–2019.
91. A. Ern and J.-L. Guermond. Finite element quasi-interpolation and best approximation. *M2AN Math. Model. Numer. Anal.*, 51(4):1367–1385, 2017.
92. A. Ern and J.-L. Guermond. *Finite Elements I: Approximation and interpolation*, volume 72 of *Texts in Applied Mathematics*. Springer, Cham, 2021.
93. A. Ern and J.-L. Guermond. *Finite Elements II: Galerkin approximation, elliptic and mixed PDEs*, volume 73 of *Texts in Applied Mathematics*. Springer, Cham, 2021.
94. A. Ern and M. Vohralík. Stable broken H^1 and $\boldsymbol{H}(\mathrm{div})$ polynomial extensions for polynomial-degree-robust potential and flux reconstruction in three space dimensions. *Math. Comp.*, 89(322):551–594, 2020.
95. A. Ern and P. Zanotti. A quasi-optimal variant of the hybrid high-order method for elliptic partial differential equations with H^{-1} loads. *IMA J. Numer. Anal.*, 40:2163–2188, 2020.
96. L. C. Evans. *Partial differential equations*, volume 19 of *Graduate Studies in Mathematics*. American Mathematical Society, Providence, RI, 1998.
97. R. Eymard, T. Gallouët, and R. Herbin. Discretization of heterogeneous and anisotropic diffusion problems on general nonconforming meshes SUSHI: a scheme using stabilization and hybrid interfaces. *IMA J. Numer. Anal.*, 30(4):1009–1043, 2010.
98. G. Fu, B. Cockburn, and H. Stolarski. Analysis of an HDG method for linear elasticity. *Internat. J. Numer. Methods Engrg.*, 102(3-4):551–575, 2015.
99. P. Grisvard. *Elliptic problems in nonsmooth domains*, volume 24 of *Monographs and Studies in Mathematics*. Pitman (Advanced Publishing Program), Boston, MA, 1985.
100. A. Grundmann and H. M. Moller. Invariant integration formulas for the n-simplex by combinatorial methods. *SIAM J. Numer. Anal.*, 15(2):282–290, 1978.
101. Q. Guan, M. Gunzburger, and W. Zhao. Weak-Galerkin finite element methods for a second-order elliptic variational inequality. *Comput. Methods Appl. Mech. Engrg.*, 337:677–688, 2018.
102. B. Halphen and Q. Son Nguyen. Sur les matériaux standard généralisés. *J. Mecanique.*, 14:39–63, 1975.
103. W. Han and B. D. Reddy. *Plasticity: Mathematical Theory and Numerical Analysis*. Springer, New York, 2013.
104. P. Hansbo. A discontinuous finite element method for elasto-plasticity. *Int. J. Numer. Meth. Biomed. Eng.*, 26(6):780–789, 2010.
105. C. Harder, D. Paredes, and F. Valentin. A family of multiscale hybrid-mixed finite element methods for the Darcy equation with rough coefficients. *J. Comput. Phys.*, 245:107–130, 2013.

106. F. Hédin, G. Pichot, and A. Ern. A hybrid high-order method for flow simulations in discrete fracture networks. In *Numerical mathematics and advanced applications—ENUMATH 2019*, Lect. Notes Comput. Sci. Eng. Springer, Cham, 2021.
107. J. S. Hesthaven and T. Warburton. *Nodal Discontinuous Galerkin Methods: Algorithms, Analysis, and Applications*. volume 54 of *Texts in Applied Mathematics*. Springer, New York, NY, 2008.
108. C. O. Horgan. Korn's inequalities and their applications in continuum mechanics. *SIAM Rev.*, 37:491–511, 1995.
109. A. Johansson and M. G. Larson. A high order discontinuous Galerkin Nitsche method for elliptic problems with fictitious boundary. *Numer. Math.*, 123(4):607–628, 2013.
110. L. John, M. Neilan, and I. Smears. Stable discontinuous Galerkin FEM without penalty parameters. In *Numerical Mathematics and Advanced Applications ENUMATH 2015*, Lecture Notes in Computational Science and Engineering, pages 165–173. Springer, Cham, 2016.
111. H. Kabaria, A. J. Lew, and B. Cockburn. A hybridizable discontinuous Galerkin formulation for non-linear elasticity. *Comput. Methods Appl. Mech. Engrg.*, 283:303–329, 2015.
112. P. Keast. Moderate-degree tetrahedral quadrature formulas. *Comput. Methods Appl. Mech. Engrg.*, 55(3):339–348, 1986.
113. N. Kikuchi and J. T. Oden. *Contact problems in elasticity: a study of variational inequalities and finite element methods*, volume 8 of *SIAM Studies in Applied Mathematics*. Society for Industrial and Applied Mathematics (SIAM), Philadelphia, PA, 1988.
114. J. Krämer, C. Wieners, B. Wohlmuth, and L. Wunderlich. A hybrid weakly nonconforming discretization for linear elasticity. *Proc. Appl. Math. Mech.*, 16(1):849–850, 2016.
115. C. Lehrenfeld. *Hybrid Discontinuous Galerkin methods for solving incompressible flow problems*. PhD thesis, Rheinisch-Westfälische Technische Hochschule (RWTH) Aachen, Germany, 2010.
116. C. Lehrenfeld and J. Schöberl. High order exactly divergence-free hybrid discontinuous Galerkin methods for unsteady incompressible flows. *Comput. Methods Appl. Mech. Engrg.*, 307:339–361, 2016.
117. S. Lemaire. Bridging the hybrid high-order and virtual element methods. *IMA J. Numer. Anal.*, 41(1):549–593, 2021.
118. J. Lemaitre and J.-L. Chaboche. *Mechanics of Solid Materials*. University Press, Cambridge, 1994.
119. K. Lipnikov and G. Manzini. A high-order mimetic method on unstructured polyhedral meshes for the diffusion equation. *J. Comput. Phys.*, 272:360–385, 2014.
120. R. Liu, M. F. Wheeler, C. N. Dawson, and R. H. Dean. A fast convergent rate preserving discontinuous Galerkin framework for rate-independent plasticity problems. *Comput. Methods Appl. Mech. Engrg.*, 199(49-52):3213–3226, 2010.
121. R. Liu, M. F. Wheeler, and I. Yotov. On the spatial formulation of discontinuous Galerkin methods for finite elastoplasticity. *Comput. Methods Appl. Mech. Engrg.*, 253:219–236, 2013.
122. W. McLean. *Strongly elliptic systems and boundary integral equations*. Cambridge University Press, Cambridge, 2000.
123. C. Miehe, N. Apel, and M. Lambrecht. Anisotropic additive plasticity in the logarithmic strain space: modular kinematic formulation and implementation based on incremental minimization principles for standard materials. *Comput. Methods Appl. Mech. Engrg.*, 191(47–48):5383–5425, 2002.
124. P. Monk and E. Süli. The adaptive computation of far-field patterns by a posteriori error estimation of linear functionals. *SIAM J. Numer. Anal.*, 36(1):251–274, 1999.
125. L. Mu, J. Wang, and X. Ye. A weak Galerkin finite element method with polynomial reduction. *J. Comput. Appl. Math.*, 285:45–58, 2015.
126. N. C. Nguyen and J. Peraire. Hybridizable discontinuous Galerkin methods for partial differential equations in continuum mechanics. *J. Comput. Phys.*, 231(18):5955–5988, 2012.
127. N. C. Nguyen, J. Peraire, and B. Cockburn. High-order implicit hybridizable discontinuous Galerkin methods for acoustics and elastodynamics. *J. Comput. Phys.*, 230(10):3695–3718, 2011.

128. J. Nitsche. Über ein Variationsprinzip zur Lösung von Dirichlet-Problemen bei Verwendung von Teilräumen, die keinen Randbedingungen unterworfen sind. *Abh. Math. Sem. Univ. Hamburg*, 36:9–15, 1971.
129. L. Noels and R. Radovitzky. A general discontinuous Galerkin method for finite hyperelasticity. Formulation and numerical applications. *Internat. J. Numer. Methods Engrg.*, 68(1):64–97, 2006.
130. R. W. Ogden. *Non-linear elastic deformations*. Dover Publications Inc., New York, NY, 1997.
131. I. Oikawa. A hybridized discontinuous Galerkin method with reduced stabilization. *J. Sci. Comput.*, 65(1):327–340, 2015.
132. L. E. Payne and H. F. Weinberger. An optimal Poincaré inequality for convex domains. *Arch. Rational Mech. Anal.*, 5:286–292, 1960.
133. M. A. Sánchez, C. Ciuca, N. C. Nguyen, J. Peraire, and B. Cockburn. Symplectic Hamiltonian HDG methods for wave propagation phenomena. *J. Comput. Phys.*, 350:951–973, 2017.
134. J. C. Simo. Algorithms for static and dynamic multiplicative plasticity that preserve the classical return mapping schemes of the infinitesimal theory. *Comput. Methods Appl. Mech. Engrg.*, 99:61–112, 1992.
135. J. C. Simo and T. J. R. Hughes. *Computational Inelasticity*. Springer, Berlin, 1998.
136. A. Sommariva and M. Vianello. Product Gauss cubature over polygons based on Green's integration formula. *BIT*, 47(2):441–453, 2007.
137. A. Sommariva and M. Vianello. Gauss-Green cubature and moment computation over arbitrary geometries. *J. Comput. Appl. Math.*, 231(2):886–896, 2009.
138. S.-C. Soon. *Hybridizable Discontinuous Galerkin Method for Solid Mechanics*. PhD thesis, University of Minnesota, MN, 2008.
139. S.-C. Soon, B. Cockburn, and H. K. Stolarski. A hybridizable discontinuous Galerkin method for linear elasticity. *Internat. J. Numer. Methods Engrg.*, 80(8):1058–1092, 2009.
140. M. Stanglmeier, N. C. Nguyen, J. Peraire, and B. Cockburn. An explicit hybridizable discontinuous Galerkin method for the acoustic wave equation. *Comput. Methods Appl. Mech. Engrg.*, 300:748–769, 2016.
141. Y. Sudhakar and W. A. Wall. Quadrature schemes for arbitrary convex/concave volumes and integration of weak form in enriched partition of unity methods. *Comput. Methods Appl. Mech. Engrg.*, 258:39–54, 2013.
142. A. ten Eyck, F. Celiker, and A. Lew. Adaptive stabilization of discontinuous Galerkin methods for nonlinear elasticity: analytical estimates. *Comput. Methods Appl. Mech. Engrg.*, 197(33-40):2989–3000, 2008.
143. A. ten Eyck, F. Celiker, and A. Lew. Adaptive stabilization of discontinuous Galerkin methods for nonlinear elasticity: motivation, formulation, and numerical examples. *Comput. Methods Appl. Mech. Engrg.*, 197(45–48):3605–3622, 2008.
144. A. ten Eyck and A. Lew. Discontinuous Galerkin methods for non-linear elasticity. *Internat. J. Numer. Methods Engrg.*, 67(9):1204–1243, 2006.
145. A. Veeser and R. Verfürth. Poincaré constants for finite element stars. *IMA J. Numer. Anal.*, 32(1):30–47, 2012.
146. C. Wang, J. Wang, R. Wang, and R. Zhang. A locking-free weak Galerkin finite element method for elasticity problems in the primal formulation. *J. Comput. Appl. Math.*, 307:346–366, 2016.
147. F. Wang and H. Wei. Virtual element method for simplified friction problem. *Appl. Math. Letters*, 85:125–131, 2018.
148. J. Wang and X. Ye. A weak Galerkin finite element method for second-order elliptic problems. *J. Comput. Appl. Math.*, 241:103–115, 2013.
149. J. Wang and X. Ye. A weak Galerkin mixed finite element method for second order elliptic problems. *Math. Comp.*, 83(289):2101–2126, 2014.
150. P. Wriggers and B. Hudobivnik. A low order virtual element formulation for finite elasto-plastic deformations. *Comput. Methods Appl. Mech. Engrg.*, 327:459–477, 2017.
151. P. Wriggers, B. D. Reddy, W. Rust, and B. Hudobivnik. Efficient virtual element formulations for compressible and incompressible finite deformations. *Comput. Mech.*, 60(2):253–268, 2017.

152. P. Wriggers, W. T. Rust, and B. D. Reddy. A virtual element method for contact. *Comput. Mech.*, 58(6):1039–1050, 2016.
153. S. Wulfinghoff, H. R. Bayat, A. Alipour, and S. Reese. A low-order locking-free hybrid discontinuous Galerkin element formulation for large deformations. *Comput. Methods Appl. Mech. Engrg.*, 323:353–372, 2017.
154. X. Ye and S. Zhang. A stabilizer-free weak Galerkin finite element method on polytopal meshes. *J. Comput. Appl. Math.*, 371:112699, 9, 2020.
155. M. Zhao, H. Wu, and C. Xiong. Error analysis of HDG approximations for elliptic variational inequality: obstacle problem. *Numer. Algorithms*, 81(2):445–463, 2019.

Printed in the United States
by Baker & Taylor Publisher Services